Springer Series in Advanced Manufacturing

Series editor

Duc Truong Pham, University of Birmingham, Birmingham, UK

The **Springer Series in Advanced Manufacturing** includes advanced textbooks, research monographs, edited works and conference proceedings covering all major subjects in the field of advanced manufacturing.
The following is a non-exclusive list of subjects relevant to the series:

1. Manufacturing processes and operations (material processing; assembly; test and inspection; packaging and shipping).
2. Manufacturing product and process design (product design; product data management; product development; manufacturing system planning).
3. Enterprise management (product life cycle management; production planning and control; quality management).

Emphasis will be placed on novel material of topical interest (for example, books on nanomanufacturing) as well as new treatments of more traditional areas.

As advanced manufacturing usually involves extensive use of information and communication technology (ICT), books dealing with advanced ICT tools for advanced manufacturing are also of interest to the Series.

Springer and Professor Pham welcome book ideas from authors. Potential authors who wish to submit a book proposal should contact Anthony Doyle, Executive Editor, Springer, e-mail: anthony.doyle@springer.com.

More information about this series at http://www.springer.com/series/7113

Mojtaba Khorram Niaki · Fabio Nonino

The Management of Additive Manufacturing

Enhancing Business Value

 Springer

Mojtaba Khorram Niaki
Sapienza University of Rome
Rome
Italy

Fabio Nonino
Sapienza University of Rome
Rome
Italy

ISSN 1860-5168   ISSN 2196-1735   (electronic)
Springer Series in Advanced Manufacturing
ISBN 978-3-319-85882-1   ISBN 978-3-319-56309-1   (eBook)
https://doi.org/10.1007/978-3-319-56309-1

Printed on acid-free paper

This Springer imprint is published by Springer Nature
The registered company is Springer International Publishing AG
The registered company address is: Gewerbestrasse 11, 6330 Cham, Switzerland

Contents

List of Exhibits

Introduction

The book resulted from a growing awareness of the ongoing mutations in many of the industrial sectors driven by Additive Manufacturing. Additive Manufacturing (AM) is the official and universal term for all applications of the technology as defined by ASTM Standard F2792: a process of joining materials to make objects from 3D model data, usually layer upon layer, as opposed to subtractive manufacturing technologies, such as traditional machining. However, AM is something more: it is a disruptive technology, and a technological shift that can lead companies into a new area of manufacturing characterized by phenomenal changes to industries and the development of new business models.

With the rapid manufacture of customized, low-cost products, AM technologies have considerable and far-reaching effects on the industrial world. They affect nearly all managerial and organizational thinking that was previously adapted to conventional manufacturing. Currently, the technology is well suited to manufacturing areas that involve high-value products with complex geometries in small and medium production volumes. It boosts the productivity of the new product development process by slashing costs, saving time and increasing creativity and innovativeness. It shrinks the supply chain by bringing firms closer to their customers.

Objective of the Book

The book aims to meet the needs of professionals and scholars; it offers the reader a view of the world of AM technologies and proposes an analysis of this theme in order to examine its attractiveness and effectiveness from a managerial point of view. The underlying questions to which the book responds are as follows:

- *How do AM technologies affect manufacturing, business strategy and business performance?*
- *How do strategic and organizational contingent factors drive AM performance?*
- *How can companies gain a competitive advantage from AM?*
- *How should companies select and implement AM technologies?*

To do this, the book aims to provide readers with a full understanding of AM technologies, their application sectors and outcomes in industrial and practical environments, and gives details on the fundamental impacts of additive technology, particularly regarding operations, innovation, supply chain, environment, and customer relationships, through scientific evidence of exemplary cases that led companies to innovative and winning business models. Moreover, it contains the results of a broad survey conducted on 105 major companies adopting AM technology.

Finally, the most important objective of the book is to provide advice on how to enhance the business value of AM technologies in different industrial and commercial environments, thereby guiding the reader to an appropriate selection and implementation of AM and related supply and manufacturing processes in different industrial environments.

Structure and Contents of the Book

The book has three sections and seven chapters visually represented in the figure below, and it contains fourteen exhibits resulting from the survey.

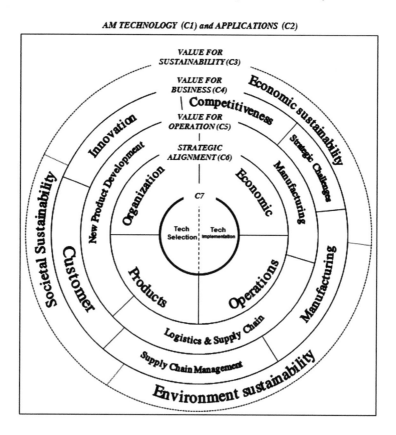

Visual Structure of the Book

The first section presents the AM technologies and their industrial applications. Chapter 1 introduces the current status of additive technologies, discusses various terminologies and outlines the historical evolution of AM technology, including earlier efforts in research and development, patents and the leading inventors and companies. Various AM technologies, and the materials available to each system along with the most widespread commercialized systems are described, including available low-cost 3D printing for both industrial grade and home systems. Chapter 2 provides details of AM application sectors. It starts with the status of the technology in the global economy and marketplace, and its role in the so-called Industry 4.0. It then discusses leading application sectors such as healthcare and the automotive and aerospace industries among others, outlining the main advantages to each industry of adopting AM technologies.

The second section of the book is at the center of our research and relates to the value of AM. Chapter 3 highlights the value of AM to sustainability from economic, environmental and societal points of view. It explains the value of AM to the economy, and it describes the positive impacts for the environment, including the impacts on energy consumption and on resource usage and pollution respectively. Finally, it discusses the social implications of AM, its advantages and the challenges that lie ahead.

Chapter 4 gives empirical evidences of the value of AM for business strategies, from its impact on a firm's competitiveness, creativity and innovation to the capability of this disruptive technology to offer new online services. Moreover, the chapter outlines key changes and their effects on operations strategies, and consequently on manufacturing and supply chain management paradigms: from mass customization to lean and agile manufacturing and supply chain management. Finally, the benefits for entrepreneurial opportunities and the role of prosumers are discussed together with the strategic and technological barriers facing the spread of AM.

Chapter 5 outlines the effects of the introduction of AM technologies on the fundamental processes in manufacturing organizations, namely new product development, manufacturing and logistics. It explains the new design methodology for AM technologies, and the exclusive benefits that AM brought to the design processes, along with providing evidence of the impacts on production process, product quality, operational costs and, as regards logistics, on inventory turnover, planning for spare-part supply chain configurations, and designing supply chain for Three Dimensional Printing (3DP) platforms.

Even though several potential impacts of AM are identified above, there is still a need to study the effectiveness of AM in different environments, industries, operation strategies, business models and processes and, in general, as a factor impacting on a company's competitive advantage. Therefore, the third section proposes a comprehensive framework for aligning companies' business strategies and operations in order to gain a competitive advantage from additive manufacturing. In fact, the positive impacts of AM on time, cost, quality, environment and business flexibility depend on specific contingent conditions.

Chapter 6 starts by proposing a framework containing some building blocks necessary for the strategic alignment of a company's business model with the adoption of AM, starting with a business strategy. Thus, the chapter gives evidences of contingency factors driving AM performances categorized by economic, organizational, operational, and product characteristics, and suggests how to perform an economic analysis to justify technology adoption. Chapter 7 starts by suggesting how to perform a technology analysis and subsequently how to perform AM technology selection and implementation, taking into account changes in the organizational structure, operations and the supply chain.

The two final chapters provide robust elements to enable managers to decide whether to adopt AM technology based on their companies' specific context and to enhance their company's competitive value through correct selection and implementation.

A final note on the empirical evidence presented in the book. We conducted a research survey by sending a questionnaire to 105 key respondents (e.g. CEO, operation managers, R&D managers, product designer, and business development managers, etc.) in companies around the world that are adopting AM technologies with the aim of identifying the real impact of AM technology on the adopters, current barriers and contingent factors driving three macro-typologies of performances: operational (time performance, cost performance, and quality performance), environmental impacts, and business flexibility. We employed five Likert scale questions in order to analyze the opinions of respondents, in which a score of five means AM is highly effective regarding the aspect in question, and a score of one means it is very ineffective. We report the analysis of responses in fourteen exhibits distributed throughout the book, and we report the analysis of contingent factors that drive AM performance in Chap. 6.

Chapter 1
What Is Additive Manufacturing?
Additive Systems, Processes and Materials

This chapter introduces the current status of additive technologies. It initially discusses various terminologies used by researchers and practitioners to define this emerging technology, in order to reach the most appropriate phrase. The origin and historical evolution of the technology are then discussed, including earlier research and development efforts, patents and the leading inventors and companies. Then, various additive manufacturing (AM) technologies and the most widespread commercialized systems are described. It provides useful information on the process of each system, its main features and application sectors, focusing on the most famous commercially available system. It first explains the liquid-based AM technologies, including stereolithography and the jetting system, then moves on to powder-based systems such as selective laser sintering, direct metal laser sintering, and electron beam melting, before finally discussing solid-based systems including fused deposition modeling and laminated object manufacturing. The materials available to each system are then discussed. The detail of available low-cost 3D printing and the top ten commercial systems for both industrial grade and home-use are then presented.

1.1 Definition

Before starting to talk about AM technologies and their value, it is necessary to decide on a single terminology with which to call this emerging technology. Many terms have been used to describe AM, which usually depicted one section of the manufacturing method, or at least did not encompass all of the applications. This was due in part to the speed of development and there is a need for a clear and standard terminology.

The term rapid prototyping (RP) was used in the industry to describe the process of rapidly creating a part before final production and commercializing. In other words, the output of this process will be a prototype or basic model. Since, the first application was only for prototyping, the term RP was used to define a process of

© Springer International Publishing AG 2018
M. Khorram Niaki and F. Nonino, *The Management of Additive Manufacturing*,
Springer Series in Advanced Manufacturing,
https://doi.org/10.1007/978-3-319-56309-1_1

layer-based fabrication. Therefore, RP means the use of layer-based techniques for producing prototypes. With the development of systems, this technique is also used for end-use parts, and users of the technology have come to realize that this term does not adequately describe some of the more recent applications of the technology. Thus, it was named as rapid manufacturing (RM). Therefore, RM means the use of layer-based techniques for producing end-usable products. Additive processes also are used in the tooling process of traditional machining or cast molding processes. These tools may include jigs and fixtures, molds and any types of complex manufacturing tools. Therefore, rapid tooling (RT) means the use of layer-based techniques in the tooling process.

In an early effort to name the technology and define its technique, Hopkinson and Dickens (2001) state that "rapid manufacture uses LMT's (Layer Manufacturing Techniques) for the direct manufacture of solid 3D products either as parts of assemblies or as stand-alone products." Rapid manufacturing (RM) is not the high-speed fabrication of parts, as its name may at first suggest, but rather refers to the use of additive technologies in the direct production of finished parts from digital data (Bak 2003). They have all the same feature of direct fabrication of the final object from 3D model data. Wohlers (2007) stated "Unlike machining processes, which are subtractive in nature, additive systems join together liquid, powder, or sheet materials to form parts. Parts that may be difficult or even impossible to fabricate by any other method can be produced by additive systems. Based on this, horizontal cross sections taken from a 3D computer model, they produce plastic, metal, ceramic, or composite parts, layer upon layer." This definition considered the different nature of this technology compared to the common subtractive manufacturing methods.

In summary, in light of the aim of using AM technology, it is known as "rapid prototyping (RP)" "rapid manufacturing (RM)" or "rapid tooling (RT)" which was briefly described above. In addition, considering the technical nature, it is also called "layered manufacturing," "additive processes," "direct digital manufacturing," "solid freeform fabrication," or "3D printing." The fundamental idea of this manufacturing method is to create a part by adding material layer by layer, in contrast to a traditional process in which we usually cut the block of material to reach the final given part, therefore the words "additive" and "layered" were used to name these methods. Moreover, since the method can fabricate a part without using any tool and mold, directly from 3D model data, the words "direct digital" and "freeform fabrication" were used to name it.

Finally, additive manufacturing (AM) is the official and universal term for all applications of the technology as defined by ASTM Standard F2792. It is defined as a process of joining materials to make objects from 3D model data, usually layer upon layer, as opposed to subtractive manufacturing technologies, such as traditional machining.

The general process of AM is clearly shown in Fig. 1.1 which depicts how a 3D object is made from 3D CAD model. The process begins with the 3D model data of the object, usually created by computer-aided design (CAD) software or a scan of an existing object. Specialized software slices this model into cross-sectional layers,

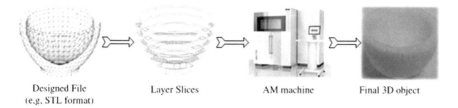

Designed File   Layer Slices   AM machine   Final 3D object
(e.g. STL format)

Fig. 1.1 General additive manufacturing process

and creating a digital file to be sent to the AM system. The AM system then generates the object by forming each layer upon another layer (Khorram Niaki and Nonino 2017a).

1.2 History

AM technology has advanced rapidly since its inception in the late 1980s. Despite this fact, it still took almost two decades of research before AM became competitive with respect to conventional manufacturing methods. The technology saw considerable technical and entrepreneurial growth over two decades. It was initially serving niche industrial design markets until the open design project and new round of startups enabled the consumer market to implement the technology.

The technology had a main application in prototyping. Prototypes allow manufacturers to evaluate a design and even to measure the performance of the products before mass production and distribution. It enables manufacturers to economically produce parts in low volumes and in less time. Thanks to this technology, prototyping that once took several months using conventional methods, was reduced to a few days or hours since it does not need the resources required by conventional manufacturing such as molds, fixtures, a long production line, and so on. In addition, it provides a freedom of design that enables designers to create parts with geometrical and structural complexity. So, AM rapidly moved to the forefront of prototyping due to these benefits along with further impacts on time, cost and quality (which will be discussed in Chap. 5).

1.2.1 Earlier Research and Development

The first effort to fabricate solid objects using photopolymer materials was in the late 1960s at Battelle Memorial Institute. DuPont invented the photopolymer resins used in the process—a type of polymer that changes its properties when exposed to light. The process involved two laser beams of different wavelength in the middle of a vat of resin, trying to solidify the material at the point of intersection. In 1967,

Wyn K. Swainson (Denmark) applied for a patent (Method of Producing a 3D Figure by Holography on a similar dual laser beam approach).

As stated by Wohlers 2014, in the early 1970s, the Formigraphic Engine Co. (founded by Swainson) employed the dual-laser approach in the first commercial laser-prototyping project. In 1974, Formigraphic presented the generation of a 3D object using a rudimentary system. In the late 1970s, Dynell Electronics Corp. assigned a series of patents on solid photography. The invention made a 3D object by the cutting of cross sections using either a milling machine or laser, and then stacking them to form the final object. Hideo Kodama from the Nagoya Municipal Industrial Research Institute was among the first inventors of the single-beam laser curing method. In 1980 he applied for a patent in Japan, which later expired before proceeding to the examination phase, which was a requirement of the Japanese patent application process. He claimed to have had difficulty in obtaining funds for additional research and development. Kodama published his second paper, titled Automatic Method for Fabricating a Three-Dimensional Plastic Model with Photo Hardening in 1981.

1.2.2   Technology Background

A series of additive technologies were invented in the twentieth century (during the late 1980s) as reported by West and Kuk (2014). Table 1.1 reports the most important patents that contributed to the development of these technologies. During that time, none emerged as a clear dominant design that displaced the others, with a market share fragmented between three or more technologies. All of these approaches include the creation of a three-dimensional object as a series of thin layers, one on top of another.

Commercially, AM first emerged in 1987 with stereolithography (SL) from 3D Systems by Chuck Hull, a process that solidifies thin layers of UV light-sensitive liquid polymer using a laser. Three AM systems were commercialized in 1991, including fused deposition modeling (FDM; from Stratasys by Scott Crump), solid ground curing (SGC; from Cubital by Itzchak Pomerantz), and laminated object manufacturing (LOM; from Helisys by Michael Feygin). The FDM process includes extruding thermoplastic materials in filament form to create parts layer by layer. SGC uses a UV-sensitive liquid polymer to solidify full layers in one pass of the UV light through masks shaped with electrostatic toner on a glass plate. The LOM process includes the bonding and cutting of sheet material using a digitally guided laser. The main and leading AM systems will be described in detail in Sect. 1.3.

Selective laser sintering (SLS) from DTM (now a part of 3D Systems) became available in 1992. Using a laser, SLS fuses powder materials. In 1994, several new AM systems were introduced. The Model Maker from Solidscape (then called Sanders Prototype) deposits wax materials using an inkjet print head. One of the new Japanese systems was a small stereolithography system (from Meiko, which

Table 1.1 Founding additive manufacturing technologies and patents

Process	First granted US patent (Date)	Key inventor (Employer)	Feedstock
Stereolithography (SLA)	4,575,330 (1984)	Chuck Hull (UVP, later 3D Systems)	Liquid plastic
Laminated object modeling (LOM)*	4,752,352 (1986)	Michael Feygin (later Helisys)	Paper
Selective laser sintering (SLS)	4,863,539 (1986)	Carl Deckard (U.Texas)	Plastic or metal powder
Solid ground curing (SGC)	4,961,154 (1986)	Itzchak Pomerantz (SciTex, later Cubital)	Liquid plastic
Fused deposition modeling (FDM)**	5,121,329 (1989)	Scott Crump (Stratasys)	Continuous spool of plastic (later metal)
Electron beam melting (EBM)	5,786,562 (1993)	Ralf Larson (Larson Brothers)	Metal powder
Inkjet-based approaches			
Three-dimensional printing (3DP)***	5,204,055 (1989)	Michael Cima, Emanuel Sachs (MIT)	Liquid plastic or plastic-metal
Inkjet printing (IJP)	5,506,607 (1991)	Royden Sanders Jr. (later Solidscape)	Wax
PolyJet	6,259,962 (1999)	Hanan Gothait (Objet)	Liquid plastic

*Trademark of Helisys
**Trademark of Stratasys
***Trademark sought by MIT, later abandoned
Source West and Kuk (2014)

ended its SL business in 2006), which was aimed mainly at the producers of jewelry. These technologies were invented either by academic researchers or by individual inventors who then went on to fund startups to commercialize the technology (Table 1.2).

3D Systems sold its first 3D printer (named the Actua 2100) in 1996, after eight years of selling stereolithography systems. The system worked with the deposition of wax material layer by layer with an inkjet printing mechanism. In March 1999, the company introduced the Thermo-Jet, which was a faster and cheaper version of the Actua 2100. A month earlier, 3D Systems launched its SLA 7000 system for around $800,000, becoming the most expensive plastic-based AM system at the time.

During this period, the price of AM systems meant it was only use by professionals. Today, developers are attempting to lower the price of desktop 3D printers so that they are affordable for the final consumers. Therefore, the growing ownership of printing machines by the final consumers is predictable. The drawing file will be downloaded and the required part will be then printed at home. Accordingly,

Table 1.2 Key AM companies

Founded	Company	HQ	Printing process	First system	Initial target	Exit
1985	**Helisys**	Los Angeles	**LOM**	1990	Industrial	1996: IPO2000: out of business
1986	**3D systems**	Los Angeles	**SLA**	1987	Industrial	1987: IPO (Vancouver)
1986	**Cubital**	Israel	**SGC**	1991	Industrial	2000: out of business
1987	**DTM**	Austin	**SLS**	1992	Industrial	2001: Acquired by 3D Systems
1989	**Stratasys**	Minneapolis	**FDM**	1992	Industrial	1994: IPO (NASDAQ)
1993	**Solidscape**	New Hampshire	**IJP**	1994	Commercial	2011: Acquired by Stratasys
1994	Z Corp	Boston	3DP	1997	Industrial	2012: Acquired by 3D Systems
1998	**Objet**	Israel	**PolyJet**	2001	Industrial	2012: Merged with Stratasys
1996**	ExOne	Pittsburgh	3DP	1999	Industrial	2013: IPO (NASDAQ)
1997	Arcam	Sweden	EBM*	2002	Industrial	2000: IPO (Nordic Growth Market)
2007***	Shapeways	Netherlands	3DP platforms	*2008*	Commercial	N/A
2009	Afinia	Minneapolis	FDM	2012	Consumer	N/A
2009	MakerBot	New York City	FDM	2009	Consumer	2013: Acquired by Stratasys
2011	RepRap	UK	FDM	2011	Consumer	N/A
2011	Ultimaker	Netherlands	FDM	2011	Consumer	N/A
2011	Formlabs	Boston	SLA	2012	Consumer	N/A

Processes invented by a company founder or employee are marked in **bold**
*Exclusive patent license from inventor
**Parent company began 3D printing in 1996, spun off in 2005
***Spun off as independent company in 2010
Source West and Kuk (2014)

Siemens predicts that 3D printing will become 50% cheaper and up to 400% faster in the next five years (Forbes 2015).

These aspects can be considered as the most important factors influencing the spread of the technology and in reducing operational costs since industrial AM machines are generally slow and expensive (Khorram Niaki and Nonino 2017b). Reportedly, AM equipment alone can account for up to half of the associated costs.

Therefore, manufacturers are seeking to increase the machinery's efficiency. For instance, they are employing multiple lasers, bigger build chambers, improved online monitoring features, and automatic changing systems, in order to develop more efficient 3D printers.

1.2.3 Evolution Phases in the Scope of AM

According to the study by Berman (2012), the scope of AM technologies has gone through three evolutionary phases in recent years (Fig. 1.2).

In the first phase, product designers employed AM technologies to produce only prototypes of new designs. AM has several key advantages in prototyping, including low cost production, reduced time to market, and privacy and security considerations. In addition, it allows a low cost modification before the final product is realizing. Therefore, AM enables the new product developer to gain several advantages over other manufacturing methods. The rapid process of producing prototypes, from several days or even weeks down to just a few hours, made the technology into mainstream prototyping and model-making tools.

Technological developments caused the second evolutionary phase, including the application of AM in creating finished parts; this step is referred to as 'direct digital manufacturing' or 'rapid tooling'. More and more manufacturers were attracted to the implementation of AM as it simplifies the supply networks, shortens lead times, and more importantly, it facilitates the innovation process needed in order to be successful in a competitive market. The main reasons for this application are the capability of the technology to produce highly complex and fully customized parts in just a few hours in a small manufacturing space. It does not need the usual, long production line with multifunctional teams and several production steps—it just needs the 3D model data and a machine to print layers of material on top of other layers. Consequently, every idea and creative design has the potential to

Fig. 1.2 The three evolutionary phases of AM scope

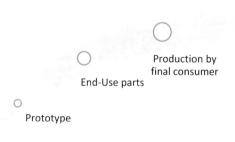

be fabricated directly by its designer. It provides, then, the chance to make much less expensive modifications, in order to obtain the optimum design and functionality. This phase, which is the application of AM in end-usable parts, has found its place in a variety of marketplaces from personal jewelry to high-tech products used in space. As the technology advances and the range of available materials increases, it is expected that we will soon see a large market of 3D printed parts, 3D printing manufacturing and specialized 3D software.

Figure 1.3 illustrates the first two phases of the evolution of AM technology accompanied with its application growth using the example of the aerospace industry. As mentioned before, applications began with prototyping and product development processes only. Then, in 2004, the aerospace industry introduced the production of components in addition to its use for prototypes. However, it was still serving very low production volumes. In 2016, General Electric (aviation) planned to mass produce 25,000 LEAP engine nozzles using AM. Therefore, it demonstrates the further development capabilities of AM technology to be a powerful competitor to conventional manufacturing processes. So, the next phase of evolution will be the technological advancements enabling larger production volumes.

Nevertheless, during the previous phases the startup cost was still expensive. The price of AM units and software was the problem and the users, therefore, were limited to professional manufacturers or designers. Efforts targeted to making low cost AM systems resulted in the third phase involving 3D printers, which, like desktop printers, are used by end consumers. This phase is the most influential and effective on consuming and manufacturing cultures as it enables consumers to produce their replacement knobs for gas ranges, chess pieces, parts for their cars, computer widgets and thousands of his/her other requirements. It provides a small factory in our homes that can produce our needs on demand, without the need for finding suppliers, paying for shipments, or losing a device due to the lack of available spare parts. However, although this application has been widely adapted for plastic material, it can be predicted for other materials considering the huge technical effort in developing printable material ranges.

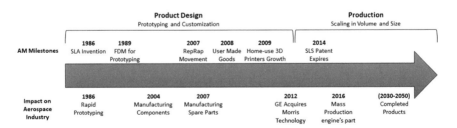

Fig. 1.3 AM technology evolution and aerospace industry. Own figure based on Cotteleer and Joyce (2014)

1.3   AM Technologies

Hopkinson and Dickens (2006) note eighteen distinct AM technologies, many of which have been commercialized in different ways by different manufacturers. Furthermore, these have been categorized based on the bulk material typology.

Table 1.3 shows the eight most widespread AM systems available. These systems will be discussed due to their dominance in AM marketplaces. The following paragraphs explain the most important AM systems, their process, advantages, and application centers.

1.3.1   Liquid-Based

1.3.1.1   Stereolithography (SLA)

SLA is the most extensive additive system in the RP process. It was the first commercialized AM process, invented by Charles Hull and introduced by 3D Systems, Inc. in 1987 (Wohlers 2014). SLA is usually used for conceptual and

Table 1.3 Available widespread AM systems

Liquid-Based	*Powder-Based*	*Solid-Based*
SLA	SLS	FDM
IJP	DMLS	LOM
	3DP	
	EBM	

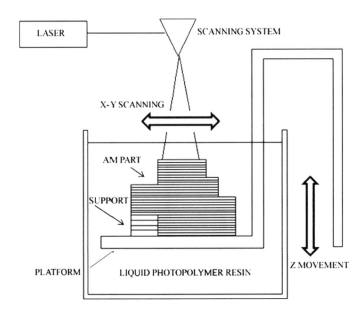

Fig. 1.4 Stereolithography (SLA) process. *Source* Monzón et al. (2015)

functional polymer prototypes. It uses an ultraviolet laser, focusing onto a photocurable liquid resin in order to build a solid part. There are more than forty
available resin types and a wide range of vendors of photopolymer resins
(Hopkinson et al. 2006). Hopkinson et al. (2006) explained the process as follows:
"using Computer Aided Design (CAD) file to drive the laser, a selected portion of
the surface of a vat of resin is cured and solidified on the platform. The platform is
then lowered, typically by 100 μm, and a fresh layer of liquid resin is deposited
over the previous layer". Figure 1.4 shows the schematic process of
stereolithography.

SLA can create very precise and detailed polymer objects, with a relatively good
surface finish (Nee et al. 2001; Mansour and Hague 2003). In addition, a wide
variety of materials and post-processing options are available for this system. SLA
is also considered as a process with a short lead time. However, the main limitation
is the requirement for supports, which need to be removed, and which consume
additional raw materials and increase the production time. In addition, the unreli-
able long-term stability of parts results in the limited application of SLA in prototyping (Petrovic et al. 2011). Another disadvantage of SLA is that the operation of
changing from one type of resin to another requires a substantial amount of time.

The main application of this system focuses on parts as master patterns (pattern
transfer process). The pattern is transferred to urethane castings, using silicone
rubber molds or is utilized for metal investment casting. In the RP process, SLA is
usually used for design appearance models, proof of concept prototypes, design

Fig. 1.5 3D printed electronic circuit board. *Image Source* Wikimedia

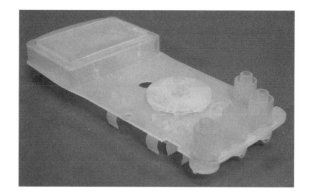

Fig. 1.6 A Stereolithography 3D printer. *Image Source* Wikimedia

evaluation models (Form & Fit), engineering proving models (Design Verification) and wind tunnel test models. In RT (rapid tooling), the process is also used for investment casting patterns, jigs and fixtures. For instance, Fig. 1.5 shows an example of a complex SLA printed electronic circuit board with various components to simulate the final product.

Figure 1.6 shows one of the SLA machines designed for desktop series. DWS (Digital Wax Systems), an Italian manufacturer of AM systems, sells an SLA 3D printer (namely the DigitalWax 030X). This machine is designed for the rapid manufacturing of industrial products with a relatively high speed of production and large product size (300 * 300 * 300 mm). Flexibility of the system is guaranteed by the wide range of available raw materials. DWS has developed its DC series of wax-based resins for direct casting and the DM/DL Series of hybrid materials for the production of master models for rubber molding applications. DM nano-filled resins are suitable for heat resistant parts with a high accuracy and excellent surface quality.

1.3.1.2 Inkjet Printing (IJP)

IJP involves the printing and curing of photocurable resins, the same as the SL process, and is typically acrylic based. There are two commercialized systems, and these are the PolyJet from Objet Systems, commercialized in 2000, and the InVision from 3D Systems, commercialized in 2003. These systems print a number of acrylic-based photopolymer material layers from printing heads containing many individual nozzles, resulting in rapid, line-wise deposition efficiency (Gibson et al. 2010). A range of about seventy materials has been introduced by Objet Geometries Ltd., with the capability of combining the materials to produce advanced composite materials and the inclusion of the properties of up to fourteen materials in a single printed model. Jetting systems can be considered as the most commonly used types of 3D printers. These are available in a range from small and inexpensive machines for consumer models, to very large and professional machines that can cost tens of thousands of dollars. Figure 1.7 shows the schematic process of Inkjet Printing.

According to Harrysson et al. (2008), IJP is suitable for RM in terms of accuracy, resolution and speed, even though material properties remain a current weaknesses of inkjet systems.

3D Systems also offers ColorJet Printing (CJP), which involves two major components of a core and binder. The system suits the building of full color concept models, architectural models and demonstration models. It also can be considered

Fig. 1.7 Schematic of PolyJet process. *Source* Bogers et al. (2016)

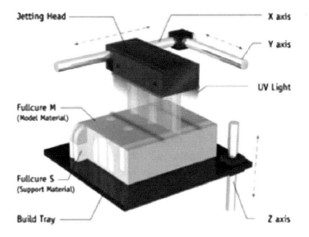

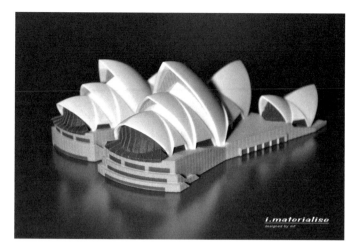

Fig. 1.8 3D printed Sydney's Opera House. *Image Source* Flicker

as a high-speed production system with a wide variety of material and colors. For instance, Fig. 1.8 shows the 3D printed Sydney's Opera House, made by *i.materialise* (an Online 3D printing service). This 3D object on a scale of 1:900 print took very slightly about few hours.

PolyJet is a great AM technology that can produce smooth and accurate parts with microscopic layer resolution and an accuracy down to 0.1 mm. It is applicable to a variety of product sizes and production volumes, from design series to manufacturing. Figure 1.9 shows one of the IJP machines of Stratasys, namely the Objet30 Prime. This is a desktop-style AM machine used for 3D printing parts by professionals or even by consumers in the home. It works with twelve different raw materials, such as rubber materials to produce gaskets or seals, and even materials for usable medical implements such as ear forms or surgical guides that require prolonged contact with the skin.

1.3.2 Powder-Based

1.3.2.1 Selective Laser Sintering (SLS)

SLS works in a similar way to SLA, but it uses powdered material as a build medium rather than liquid. Carl Deckard first developed selective laser sintering (SLS) at the University of Texas, USA. It was developed and commercialized by DTM (acquired by 3D Systems) during the early 1990s (Wohlers 2014). SLS involves the sintering of powder materials using a laser. This system is being employed as a manufacturing method for end-usable products through generating point-by-point and layer-by-layer parts.

SLS has also been introduced as a process of creating functional metal parts with accurate dimensional geometries, thus it has a significant advantage over conventional RP techniques (Liu and Li 2004). In theory, SLS can sinter any material that can be made available in powder form (Kruth et al. 2003), but currently and practically this is not the case (Goodridge et al. 2012). A variety of raw materials including polymers, ceramics and metals is currently available for the SLS process. Hopkinson et al. (2006) described the process as follows: "the powdered raw material is sintered or melted by a laser that selectively scans the surface of a powder bed to create a two-dimensional solid shape. A fresh layer of powder, typically 100 μm thick, is then added to the top of the bed so that the subsequent two-dimensional profile can be traced by the laser, bonding it to the layer below. The process continues to create a full three-dimensional object." The unfused powder material supports the part during the process. Figure 1.10 shows the schematic process of selective laser sintering.

Powder-based AM systems are more suitable for manufacturing than liquid-based processes in the case of long production runs due to its material properties and the stability of parts made by powder-based AM systems, like SLS (Hopkinson et al. 2006). However, the particle size and morphology affect both the processing ability and sinterability, resulting in quality aspects such as surface roughness, part density, accuracy and porosity among others (Shi et al. 2004; Goodridge et al. 2006, 2012). The unfused powders act as a support structure for the model, and then, at the end of the production, unprocessed powders are either disposed of or recycled, leading to lower material usage. In addition, the unprocessed powders in this system are generally less hazardous than SLA liquids. Finally, it also needs post-processing. A variety of finishing processes can be applied, including manual/tumble grinding, painting or coating by direct application on the sandblasted part, and vacuum metalizing.

SLS is a 3D printing choice for a range of functional and durable prototype applications, such as those with snap fits and mechanical joints. The ability of SLS to simultaneously print several pieces also makes it a good choice for the direct digital manufacturing (DDM) of products demanding strength and heat resistance. The range of materials which are commercially available includes aluminum-filled

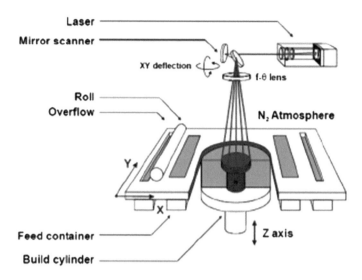

Fig. 1.10 Selective laser sintering (SLS) process. *Source* Deckers et al. (2012)

(PA12-AL), impact-resistant nylon (Duraform EX), nylon (Duraform PA), glass-filled nylon (Duraform GF), and rubber-like (Duraform Flex Plastic).

In the RM process, SLS is applied in a variety of industries such as aerospace, medical and healthcare, electronics, packaging, connectors, and the military, etc. In the RP process, it can also be useful for functional proof of concept prototypes, product performance and testing, engineering design verification, wind tunnel test models and other areas of prototyping. In the RT process, SLS can be used for injection mold inserts, tooling estimating visual aid, investment casting patterns, jigs and fixtures, and foundry patterns (sand casting). Among thousands of case studies, Fig. 1.11 illustrates a unique SLS printed running shoe midsole customized to an individual's foot, named Futurecraft 3D. Adidas aims at linking the existing data sourcing and foot-scan technologies, it provides unique opportunities for immediate in-store fittings. The midsole is made of thermoplastic polyurethane (TPU), which is durable fully-flexible AM material to be used in consumer products.

There are many commercialized SLS printing machines available worldwide. EOS is one of the world's leading 3D printing manufacturers, and was founded in 1989 in Germany. The company commercialized its first laser sintering system in 1994, namely the EOSINT P350, as the first European laser sintering system for plastic prototyping. From that date to now, its laser systems have fundamentally improved. Currently, the company offers an advanced SLS system, namely EOS P 770, a productive 3D printing system for product lengths of up to one meter and with a build volume of more than 150 L. Ten commercial polymer materials and more than eighteen potential combinations demonstrate its excellent flexibility. Fig. 1.12 shows an SLS 3D printer.

Fig. 1.11 Adidas Futurecraft personalized shoe. *Image Source* Flicker

Fig. 1.12 EOS FORMIGA 110, laser sintering systems. *Image Source* Flicker

1.3.2.2 Direct Metal Laser Sintering (DMLS)

DMLS is another commercial technique among laser-based additive technologies similar to SLS. DMLS was developed by EOS in 1995 and involves metallic powder only, while SLS is able to process a variety of materials. The difference between DMLS and SLS includes the production of metal parts without the need for a binder coating and the subsequent processing that would be required otherwise. Some of the most commonly used metals include cobalt chromium, titanium alloys, steel alloys and tool steels. The initial aim of developing DMLS was to

Fig. 1.13 Direct metal laser sintering (DMLS) process. *Source* Singh et al. (2017)

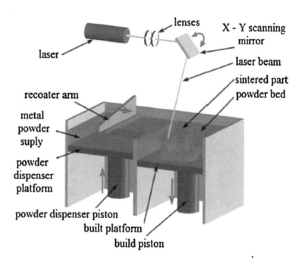

produce tooling, but the process has also been used for the production of end-use parts (Hopkinson et al. 2006). Figure 1.13 shows the schematic process of DMLS.

Mumtaz and Hopkinson (2010) demonstrated that properties can be compared with those of cast or machined components. Support structures are required for overhanging features and anchors are required due to the high thermal stresses involved in the process. In a similar way to SL, these support structures require more overall material and post-processing. This exciting process allows the production of small metal parts with extremely complex geometries that cannot be made using conventional technologies. Production can be anything from prototypes to larger production volumes of up to 20,000 units. It is an ideal manufacturing method for use in the aerospace and defense industries, although it can also be applied in machinery, tooling, food and pharmaceutics, petrochemical, automotive, veterinary and medical devices, air/oil/fuel mixing devices, sporting goods, industrial burner parts, radiation collimators, and customized production line equipment. The European Space Agency in collaboration with 3D Systems, used DMLS machines to produce injectors, combustion chambers and expansion nozzles of a bi-propellant communication satellite engine. In the new product development process, they were able to make valuable changes enabling them to optimize the design by reducing the weight of parts, simplifying assembly, increasing the speed of manufacturing, and supporting late-stage design adaptations. Fig. 1.14 shows DMLS printed metallic space fabrics, made by NASA. It is 3D printed in one piece with sintering metals layer upon layer. The fabrics can have huge applications in space, including large antennas because its shape can change quickly. The fabrics can be used to shield a spacecraft from meteorites, for astronaut spacesuits, or for capturing objects on the surface of another planet. Another potential application can be for an icy moon like Jupiter's Europa, where these fabrics could insulate the spacecraft. AM could add functionalities to the material for this complex object at less operational costs, which the engineers at NASA call it mass production of functions instead of traditional mass production.

Fig. 1.14 Metallic space fabrics. *Image Source* NASA/Jet Propulsion Laboratory

A new DMLS machine, the EOS M 400-4, was designed and manufactured by EOS for use in industrial applications. It has four 400 W lasers generating about four times more productivity in operation, a relatively large build volume of 400 * 400 * 400 mm, and a high speed of production at approximately 100 cm^3/h. A wide range of materials are available for this kind of machine, from light metals and stainless steel to super alloys.

1.3.2.3 Three Dimensional Printing (3DP)

Michael Cima and Emanuel Sachs invented 3DP in 1989 at the Massachusetts Institute of Technology (MIT). Since then, 3DP has been licensed to many companies. Z-Corp (acquired by 3D Systems in 2012) is the most important manufacturer and developer of 3D printers. The process is very similar to the SLS process, except a binder is used for the solidification of the powder instead of a laser. This process uses an inject-printing head to eject a bonding material onto successive layers of powder. The material can be any powder-based material, including plastics, metals, ceramics, or cermet (Mansour and Hague 2003). At the end of the production process, loose powder attached to the model is removed by a pneumatic system that can be used again in subsequent production. The part may then be post-processed in order to improve the surface finish and strength through infiltration with secondary resins, which fills any voids in the part. Figure 1.15 shows the schematic process of three dimensional printing.

1.3.2.4 Electron Beam Melting (EBM)

The EBM process was first commercialized by Arcam (Sweden) in 1997, and is similar to other powder-based processes in that the powder material is fused together selectively. EBM works with a thermionic emission gun, using a tungsten

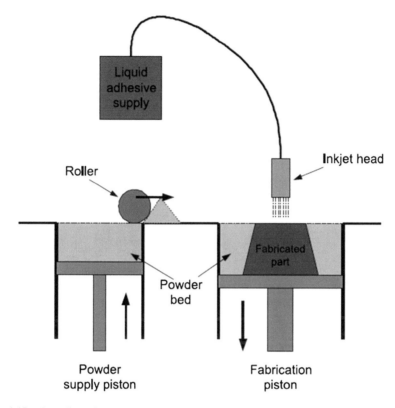

Fig. 1.15 Three dimensional printing (3DP) process. *Source* Liu et al. (2006)

filament to make an electron beam. The process selectively melts metal powder in layers of 70–250 micron thickness, with each layer being preheated by scanning the beam to lightly sinter the particles (Harrysson et al. 2008). EBM machines utilize a high power electron beam that generates the energy needed for a high melting capacity and high productivity. Electromagnetic coils provide very fast and accurate beam control, allowing several melt pools to be maintained simultaneously. This sintered powder surrounding the part helps support downward facing surfaces and breaks up during post-processing, allowing for most of the unmelted powder to be recovered and reused.

The advantages over laser-based processes (Hopkinson et al. 2006) are an increased scanning speed resulting in a reduced build time and reduced thermal stresses during the process due to the scanning technique. However, the material ranges are limited to conductive metal powders and there is a relatively poor surface finish in comparison with a laser process. The main advantage of this process is the use of the vacuum chamber, which facilitates an optimal manufacturing environment for oxygen reactive materials. Material impurities due to oxygen are strictly prohibited for safety reasons, these issues being most significant in the production of medical implants and aerospace components. Figure 1.16 shows the schematic

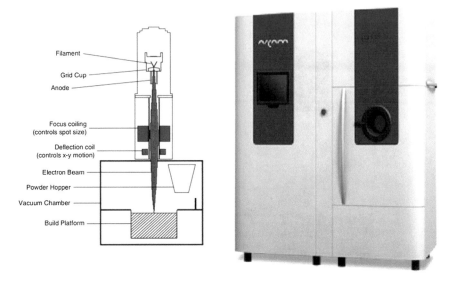

Fig. 1.16 Schematic of Electron beam melting (EBM) process, and Arcam Q10 Plus. *Image Source* ARCAM AB

process of three dimensional printing. The Arcam Q10 plus is the newest EBM machine and is specifically designed for orthopedic implants. Another of this manufacturer's EBM systems is designed to process titanium alloys offering a build envelope of 200 * 200 * 380 mm, which is a productive machine for industrial sectors, particularly the aerospace industry.

1.3.3   Solid-Based

1.3.3.1   Fused Deposition Modeling (FDM)

FDM is the second most extensively established AM system in RP after SLA. It was invented by S. Scott Crump in Eden Prairie, Minnesota, and was commercialized and introduced by Stratasys in 1991 (Wohlers 2014). FDM is one of the most important AM systems due to its ease of operation, low cost of machinery, high modulus of part made by the process, and easy material changeability (Rosochowski and Matuszak 2000; Levy et al. 2003). It is mostly used in prototyping, modelling, and manufacturing applications. The materials available for this kind of system include investment casting wax, polycarbonate, polyphenylsulfone (PPSF), and most commonly, acrylonitrile butadiene styrene (ABS) (Chua et al. 1998; Hopkinson et al. 2006), with different properties appropriate to the production of functional parts. Hopkinson et al. (2006) described the process as follows: "FDM creates parts by extruding material (normally a thermoplastic polymer)

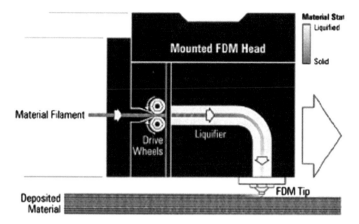

Fig. 1.17   Fused deposition modelling (FDM) process. *Source* Ahn et al. (2002)

through a nozzle that traverses in X and T to create each two-dimensional layer. In each layer separate nozzles extrude and deposit material that forms the parts and material that form supports where required." Figure 1.17 shows the schematic process of fused deposition modelling.

As mentioned above, the process is very easy to operate and is clean enough to be used in an office environment. In addition, the low-cost of the machine means the system can be applied widely in different fields. The support structure is usually made from different materials from that used for the parts and allows overhang. The support structure, when soluble supports are employed, can simply be dissolved, resulting in the production of more complex geometries. On the other hand, using a nozzle with a relatively large diameter limits the resolution and accuracy, though recently launched machines are capable of producing much smoother surfaces. Since the nozzles must physically move above the object (rather than lasers), the process speed is relatively low. The uneven heating and cooling cycles due to the inherent nature of the process cause stress accumulation in the built part resulting in distortion which is primarily responsible for weak bonding and thus affects the component strength (Anitha et al. 2001; Mansour and Hague 2003; Wang and Shaw 2005; Hopkinson et al. 2006).

Manufacturers in the medical device, aerospace, automotive, and other industries rely on FDM capabilities for the rapid supply of small volume production of consumer products, concept models, functional testing, and initial prototypes. These parts have accurate dimensions and functionality, and are able to resist high temperatures, mechanical stresses and chemical degradation. High strength, a wide variety of available materials, and a high manufacturing speed are the main features of the FDM process. Figure 1.18 shows a sample 3D printed part using an FDM system.

Due to its simpler machinery and production mechanism, many manufacturers are able to produce FDM 3D printers, and there are numerous suppliers of the system. Figure 1.19 shows one of the hundreds of these 3D printers.

Fig. 1.18 A sample printed product using FDM

Fig. 1.19 FDM system.
Image Source Wikimedia

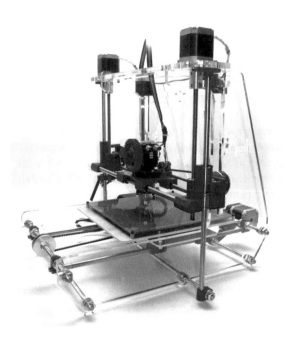

1.3.3.2 Laminated Object Manufacturing (LOM)

Michael Feygin invented the LOM system in 1985 and Helisys commercialized the process in 1991. LOM is the part of sheet stacking technologies and the process involves the stacking together of layers of material cut via laser binding to create a part. The material in LOM is a special kind of paper that has a heat-sensitive adhesive applied to one side. A number of materials are available for use in the process, including polyvinyl chloride (PVC). Hopkinson et al. (2006) described the process as follows: "LOM involves stacking layers of paper with a bonding material and creating the part profile by cutting each layer of paper with a laser. Post processing involves using hand tools to remove the unwanted material and to reveal the part inside." Figure 1.20 shows the schematic process of laminated object manufacturing.

The LOM system has been employed for both RP and RM. However, the system has fewer applications in the manufacture of end-use parts. Mueller and Kochan (1999) stated some reason for the reduced applicability of the process, including the limited part accuracy due to the comparably simple machine design. Also as with other AM systems, mechanical and thermal material properties are inhomogeneous and the detail reproduction and durability of small part features is comparably low. However, the main problem relates to the complex geometries. For instance, for those products with thin walls, post-processing is difficult, time consuming and can damage the part. However, this is not the case for simple geometries. SD 300 3D printer made by Solidimension Ltd., is a desktop LOM machine that actually fits on a desk. Early stage concept modeling, design iteration and sharing of the design

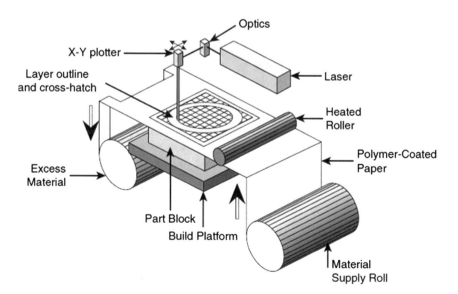

Fig. 1.20 Laminated object manufacturing (LOM) process. *Source* Gibson et al. (2010)

with professionals and customers alike are all possible to users of this low cost printer. Its maximum build size is 160 * 210 * 135 mm with a minimum wall thickness of 1 mm.

In addition to the classification of bulk material typology, an overview is proposed in Table 1.4. Laser-based AM includes two subcategories—laser melting and laser polymerization. The former includes SLS and DMLS, while the latter includes SLA. The thermal extrusion method includes FDM, while LOM is in the category of material adhesion. In addition, IJP and 3DP constitute the material jetting subgroup. In addition, the last column shows whether the system needs a support structure for the adding of layers.

1.4 Methods of fabrication and Materials

As mentioned earlier, AM technologies can be also categorized in terms of the method of fabrication. Categories include vat photopolymerization, powder bed diffusion, extrusion-based process, material jetting, binder jetting and sheet lamination processes. This section discusses the available and commercialized raw materials for use in these manufacturing processes.

1.4.1 Vat Photopolymerization

Photopolymerization processes use liquid-based curable resins as a source of material. SLA is a type of photo polymerization and is described in Sect. 1.3.1. The first patents of SLA resins were published in 1989 and 1990 (Murphy et al. 1989; Hull 1990). These resins were made from acrylate, which had high reactivity but typically produced weak parts due to the inaccuracy caused by shrinkage and curing. Other patented resins for the SLA process were epoxy-based resins. These produced more accurate, harder, and stronger parts than the acrylate resins (Petrovic et al. 2011). However, the epoxy resins have the disadvantages of a slow photo speed and brittleness of the cured parts. Therefore, some acrylate has been added to the epoxy resins, and most of the current resins available to the SLA process are epoxides with acrylate content (Gibson et al. 2010). It actually combined the advantages of both curing types.

1.4.2 Powder Bed Fusion

Any material that can be melted and re-solidified, can be used in this process. This includes polymers, metals and ceramics for use in a variety of AM processes such as SLS, SLM, and DMLS. Currently, the most common polymer material used in

Table 1.4 AM technologies overview

Manufacturing method	AM system	Bulk material type			Materials							Support Structure Required
		Liquid	Powder	Solid	Organic/wax	Paper	Ceramic	Photopolymers	Polymers	Metal	Sand	
Laser polymerization	SLA	×						×				Yes
Material jetting	IJP	×			×			×				Yes
	3DP	×	×				×	×		×		No
Laser/electron beam melting	SLS		×				×	×	×	×	×	
	DMLS		×							×		
	EBM		×							×		
Extrusion thermal	FDM			×						×		
Material adhesion	LOM			×		×						Full block used

these processes is the polyamide—a thermoplastic polymer, commonly known also as nylon. Elastomeric thermoplastic polymers are also available for building highly flexible parts with rubber-like characteristics, which are used in the production of gaskets, industrial seals, shoe soles, and other components. Several types of bio-compatible and biogradable polymers such as polycaprolactone (PCL) and poly-lactide (PLA) have been processed (Gibson et al. 2010). EOS markets a variety of materials based on PA 12 or polystyrene for use in their laser sintering systems (Petrovic et al. 2011). Furthermore, a variety of metals has been used in the powder bed fusion processes, with several types of steels, titanium and its alloys, nickel-base alloys, some aluminum alloys and cobalt-chrome all commercially available. Additionally, there are some precious metals such as silver and gold. Several ceramic materials such as aluminum oxide, titanium oxide and cermet, which are metal-ceramic composites, are also commercially available. Biocompatible materials have been also commercialized for specific applications. For instance, calcium hydroxyapatite, a material very similar to human bone, has been processed for use in medical applications.

1.4.3 Extrusion-Based

FDM is one of the more important extrusion-based processes. Currently, the most popular material is ABS-plus, which is an updated version of the original acry-lonitrile butadiene styrene (ABS) (Gibson et al. 2010). These materials have properties that are similar to thermoplastic material for injection molding. In addition, nylon-based materials and other polymers can be used. The main advantage of FDM technology is in the use of low-cost materials. Regarding available materials, ABS is suitable for functional prototype construction (Petrovic et al. 2011). There are some other materials that can be used instead of ABS in situations where ABS cannot fulfill the requirements. A material, usually poly-carbonate based, offers higher tensile properties, and in the case when the object needs heat deflection, polyphenylsulfone (PPSF) would be also an option.

1.4.4 Material Jetting and Binder Jetting

IJP is the best known of the material jetting systems. Polymers, metals and ceramics constitute the available types of materials, and the maximum printable viscosity threshold is generally considered to be in the range of 20–40 centipois at the printing temperature (Derby and Reis 2003). To facilitate the use of jetting mate-rials, they must be heated so that they liquefy. For high viscosity fluids, the vis-cosity of the fluid must be lowered to enable jetting. The limitation on viscosity became the most problematic issue for droplet formation in material jetting.

3DP is the original name for the binder jetting system. 3D Systems supplies plaster-based powder with the water-based binder. The printed parts are usually weak, and are also usually infiltrated with another material. 3D systems, for instance, provides three infiltrants, namely ColorBond infiltrant, which is acrylate based, StrengthMax infiltrant, which is a two-part infiltrant, and Salt Water Cure, which is an eco-friendly and hazard-free infiltrant. VoxelJet also supplies a PMMA (polymethyl methacrylate) powder. ExOne supplies more than 3100 stainless steel and bronze in metal powder. For stainless steel materials, bronze is used as an infiltrant. Both companies also provide sand for the fabrication of molds and cores for sand casting. ExOne supplies a silica sand two-part binder as well as a soda-lime glass material applicable in artwork, jewelry, or other decorative objects, while VoxelJet also offers a silica sand inorganic binder and claims that their material also integrates well into existing foundry processes.

1.4.5 Sheet Lamination Process

LOM is an example of a sheet lamination process and was previously described in Sect. 1.3.3. A variety of materials including plastics, metals, ceramics and paper is available for this process. Butcher paper, coated with a thin layer of thermoplastic polymer, was the first material used in lamination processes. In addition, Mcor Technologies developed a printer that uses a standard copy of paper, for instance A4 size. The water-based glue then binds the paper sheets, resulting in rigid parts. In metals, aluminum and low-carbon steel are the most commonly used materials. In addition, with ceramics, tap casting methods form sheets of material composed of powdered ceramics, such as SiC, TiC-Ni composite, or alumina, with a polymer binder (Gibson et al. 2010).

Table 1.5 details both the commercialized AM polymers, metals, ceramics, and composites and those still under-development.

Table 1.5 Commercial and upcoming AM materials and corresponding systems

Material type		AM system	Material(s)
Polymers (a)	Thermo-setting	SLA, IJP	Photocurable polymers
	Thermo-plastic	IJP	Wax
		SLS	Polyamide 12, GF polyamide, polystyrene
		FDM	ABS, PC-ABS, PC, ULTEM
		3DP	Acrylic plastics, wax

(continued)

Table 1.5 (continued)

Material type		AM system	Material(s)
Metals (a)		SLM/ DMLS	Stainless steel GP1, PH1 and 17-4, cobalt chrome MP1, titanium Ti6Al4 V, Ti6Al4 V ELI and TiCP, IN718, maraging steel MS1, AlSi20 Mg
		EBM	Ti6Al4 V, Ti6Al4 V ELI, cobalt chrome
Ceramics (b)		SLA	Suspension of Zirconia, silica, alumina, or other ceramic particles in liquid resin
		FDM	Alumina, PZT, Si3N4, zirconia, silica, bioceramic
		SLS	Alumina, silica, zirconia, ZrB2, bioceramic, graphite, bioglass, and various sands
		3DP	Zirconia, silica, alumina, Ti3SiC2, bioceramic, and various sands
Composites (b)		FDM	Polymer-metal, polymer-ceramic, short fiber-reinforced composites
		3DP	Polymer-matrix, metal-ceramic, ceramic-ceramic short fiber-reinforced composites
		LOM	Polymer-matrix, ceramic-matrix, fiber and particulate reinforced composites
		SLS, SLM	Metal-metal, metal-ceramic, ceramic-ceramic, polymermatrix, short fiber-reinforced composites

Notes (a) Commercially available materials; (b) materials under research and development
Adapted from Guo and Leu (2013)

1.5   Low-Cost Desktop 3D Printing

As will be discussed in Chap. 4, 3D printing can drive new business models and can elevate the role of consumers towards real entrepreneurial activities, but the crucial first step is its adoption by consumers. In fact, the main barrier to the widespread adoption and implementation of AM technologies as they relate to consumer goods is the matter of affordable machine cost. This is the case with respect to household prosumption that could produce a big shift in our lives. With a wider distribution of the technology in the future, the consuming pattern will change, and everyone will be able to produce at home whatever he or she wants in a minimum of time and at an even cheaper price than mass production currently offers. Any creative idea will be physically achievable if the affordable AM machines follow the path of 2D printers.

As mentioned in Sect. 1.3, different technologies exist for additive process, so the question is which one can meet the requirements of household 3D printers and domestic markets? The first criterion is definitely the lowering of cost to place it within a household budget. An acceptable level of performance and quality of 3D printed parts is the second important parameter in selecting technology for the prosumer marketplace. Seeing that AM technology should work in the house or the office, spatial efficiency and zero emissions should be considered as important factors. A summary of the criteria in selecting the most appropriate technologies for a home-use machine are listed:

- Lower cost (matched to a household budget)
- Suitable performance (suitable for the manufacture of consumer items)
- Longitudinal efficiency (should fit into a home)
- Zero emissions (to eliminate any extra installation infrastructure)
- Simple material handling (for ease of use)
- Zero post-processing would be ideal (so that the product is immediately ready with minimum effort on the part of the owner)
- Simpler process (no high skill requirement)
- Higher versatility (to provide flexible manufacturing).

An example of this evaluation can be seen in Table 1.6, in which three AM systems are compared in terms of the above-mentioned criteria. It shows that FDM and LOM meet all three major areas, while 3DP still has some challenges to overcome before it is suitable.

A further study ranked the technologies for the 3D printing of consumer goods. Bogers et al. (2016) carried out the research, adding other criteria such as time and cost of production, size of the build chamber, and use of multicolor and decoration. Multicolor refers to the ability of the AM technology to make a part in a single color without post-processing, while decoration means the ability to create colored patterns on the part.

As seen in Table 1.7, some technologies show greater potential due to their inherent characteristics. Fused deposition modeling (FDM), for example, has a higher score than others relating to its use for consumer goods and by prosumers. However, progress in other technologies, for instance in biocompatible materials for ink-jet printing (IJP) and color technology for selective laser sintering (SLS), may change this rank in the future.

Table 1.6 Evaluation of AM technologies for use in the home

AM systems	Emissions	Material handling	Post-processing
FDM	Suitable: zero emissions	Suitable: filament is contained in reels	Suitable: not required
LOM	Suitable: minor solvent emissions	Suitable: plastic sheet is contained in reels	Suitable: not required
3DP	Suitable: zero emissions	Unsuitable: starch or plaster powder is hard to completely contain	Unsuitable: required

Table 1.7 Ranking of AM technologies for consumer goods

Decision criteria	AM technologies			
	FDM	SLS	SLA	IJP
Mechanical strength	3	3	2	3
Chemical properties	5	5	1	2
Surface finish	2	3	5	4
Cost	5	2	1	1
Time	1	1	1	1
Build chamber size	2	2	2	2
Multicolor	4	2	1	4
Decoration	1	1	1	4
Sum of scores	23	19	14	21

Source Bogers et al. (2016)

A leading project in the supply of affordable 3D printing machines is the RepRap (replicating rapid prototyper) home-use 3D printer, which has grown rapidly in popularity in recent years. Self-replication refers to the process of making a copy of itself. The idea that 3D machines should be affordable to the mainstream market and should be owned and used by final consumers at home was initiated by Adrian Bowyer in 2005. Although at the beginning of this project the number of adopters was very low but by the midpoint of 2010 the RepRap community included more than 3800 individuals, and currently its users own more than 40,000 3D printers.

This AM machine uses the FDM process to fabricate a set of most of its own parts. The RepRap uses a heated nozzle to extrude the fine filament or molten plastic. Using a computer-driven motor, the nozzle moves in X, Y and Z dimensions to form the part. These are then assembled with standard, readily available components into a working copy of the original machine, with a kit that one can assemble with a small investment, typically of about $400. This machine is then capable of making the parts for future copies, and so on.

Once a person has a copy of the machine, he or she may then connect it to a computer and manufacture anything within the bounds of the FDM process. Figure 1.21 illustrates the concept for use in the home.

In recent years, user-founded companies (members of the RepRap community) have begun to sell the fully assembled home-use 3DP, instead of offering kits that users need to assemble themselves. Examples of these companies include Bits From Bytes (United Kingdom; then acquired by 3D Systems); Makerbot Industries (USA); and Ultimaker Ltd. (Netherland). These 3D printers serve the low-end market segment such as individual designers, students, inventors or artists.

The RepRap printer offers a radical alternative to the way our society manufactures and consumes. It also offers distributed manufacturing instead of current centralized manufacturing. Centralized manufacturing refers to the mass production at one site, and subsequent transporting of goods to markets. Distributed manufacturing is the production of few goods at the location of market, therefore the limited transportation usually for raw material is required. Industrial examples of

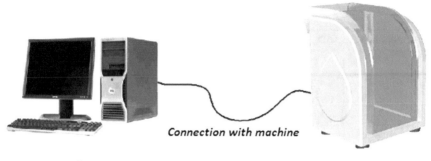

Fig. 1.21 Schematic of using RepRap printer in the home

distributed mechanical manufacture exist to meet extremely personal demands and one example of this is 2D printing. With a word processor and a printer, you can print your own documents and photographs. With a suitable kitchen, you can also cook your meals—resources and skills permitting.

1.6 Top Ten Commercial AM Systems

Along with the above-mentioned advantages of each AM technology, there are several commercialized AM machines available for each AM technology. By considering the quality and reliability of each machine, a general ranking of them is possible and these selection guidelines may help beginners and intermediate users of additive systems to select the best option. One of these rankings is the evaluation conducted by 3D Hubs.

3D Hubs is a leading 3D printing network, with more than 6900 members from over 150 countries. It provides online facilities using which anyone can upload a 3D design, select a material, and choose a local 3D printing service (members). In January 2017, the company ranked the available desktop 3D printers and industrial grade systems based on the reviews of AM users. Table 1.8 details the top ten commercialized desktop 3D printers and some of their technical features including build size and layer thickness. The respondents to this survey were asked to rate these 3D printers (from 1 to 5) based on several parameters such as print quality, ease of use, machine reliability, failure rate, customer service, open source system community, operating expenses, and available software. The current price of each system is also listed in the table. In addition, Table 1.9 shows the top ten industrial grade AM machines, their prices and technical features. Stratasys and 3D Systems are the two leading AM equipment suppliers, owning the largest market share in the world.

Table 1.8 Top ten desktop (home-use) 3D printers (own table based on the information of 3D Hubs, 2017)

	Printer model	Tech	Build size (cm)	Layer thickness (micron)	Price (USD)	Print quality rating
1	Original Prusa i3 MK2	FDM	25 × 21 × 20	50	699	4.87
2	BCN3D SIGMA	FDM	21 × 29.7 × 21	50	2795	4.85
3	Form 2	SLA	14.5 × 14.5 × 17.5	25	3925	4.83
4	PowerSpec 3D Pro	FDM	22.6 × 14.5 × 15	100	999	4.82
5	ORD Bart Hadron	FDM	19 × 19 × 15	100	699	4.8
6	Zortrax M200	FDM	20 × 20 × 18.5	90	1650	4.8
7	Kossel	FDM	26 × 26 × 27.5	30	329	4.79
8	LulzBot Mini	FDM	15.2 × 15.2 × 15.8	50	1250	4.79
9	FlashForge Creator Pro	FDM	22.5 × 14.5 × 15	100	899	4.78
10	CEL Robox	FDM	21 × 15 × 10	20	999	4.78

Seven Key Facts

- Additive Manufacturing (AM) is the official and universal term for a set of additive technologies.
- AM first emerged commercially in 1987 with stereolithography (SLA) from 3D Systems Inc. by Chuck Hull.
- AM technology has seen three phases of evolution: from application in prototyping only, to the manufacturing of end-usable products, and production with home-use desktop 3D printing.
- AM technologies are categorized based on the typology of raw material, such as liquid-based, powder-based, and solid-based.
- A range of polymers, metals, ceramics, and composites are currently available for different AM technologies, and many more are under development.
- Adrian Bowyer initiated the RepRap project in 2005. This is a 3D printer affordable to the mainstream market.
- Fused deposition modeling, laminated object manufacturing, and three dimensional printing are the AM systems that can meet the requirements of household 3D printers and domestic markets.

Table 1.9 Top ten industrial 3D Printing (own table based on the information of 3D Hubs 2017)

	Manufacturer	Printer model	Tech	Build size (cm)	Layer thickness (micron)	Price (USD)	Print quality rating
1	3D systems	Vanguard	SLS	37 × 32 × 44.5	100	–	4.97
2	3D systems	Projet 3500 HDMax	Jetting	29.8 × 18.5 × 20.3	16	70,000	4.95
3	Stratasys	Objet260 Connex	PolyJet	25.5 × 25.2 × 20	16	120,000	4.9
4	Stratasys	Objet Eden260	PolyJet	25.5 × 25.2 × 20	16	19,800	4.86
5	EOS	EOSINT P760	EBM			–	4.85
6	Stratasys	Objet Alaris30	PolyJet	30 × 20 × 15	100	24,900	4.85
7	Stratasys	Dimension 1200es	FDM	25.4 × 25.4 × 30.5	100	18,900	4.85
8	Stratasys	Objet30 Prime	PolyJet	29.4 × 19.2 × 14.86	28	30,000	4.84
9	3D systems	ProJet CJP 660Pro	Jetting	25.4 × 38.1 × 20.3	100	59,650	4.84
10	(acquired by) 3D systems	Zcorp	Binder jetting	25.4 × 38.1 × 20.3	100	999	4.83

References

Ahn, S. H., Montero, M., Odell, D., Roundy, S., & Wright, P. K. (2002). Anisotropic material properties of fused deposition modeling ABS. *Rapid prototyping Journal, 8*(4), 248–257.

Anitha, R., Arunachalam, S., & Radhakrishnan, P. (2001). Critical parameters influencing the quality of prototypes in fused deposition modelling. *Journal of Materials Processing Technology, 118*(1), 385–388.

Bak, D. (2003). Rapid prototyping or rapid production? 3D printing processes move industry towards the latter. *Assembly Automation, 23*(4), 340–345.

Berman, B. (2012). 3-D printing: The new industrial revolution. *Business horizons, 55*(2), 155–162.

Bogers, M., Hadar, R., & Bilberg, A. (2016). Additive manufacturing for consumer-centric business models: Implications for supply chains in consumer goods manufacturing. *Technological Forecasting and Social Change, 102,* 225–239.

Chua, C. K., Chou, S. M., & Wong, T. S. (1998). A study of the state-of-the-art rapid prototyping technologies. *The International Journal of Advanced Manufacturing Technology, 14*(2), 146–152.

Cotteleer, M., & Joyce, J. (2014). 3D opportunity: Additive manufacturing paths to performance, innovation, and growth. *Deloitte Review, 14,* 5–19.

Deckers, J., Shahzad, K., Vleugels, J., & Kruth, J. P. (2012). Isostatic pressing assisted indirect selective laser sintering of alumina components. *Rapid Prototyping Journal, 18*(5), 409–419.

Derby, B., & Reis, N. (2003). Inkjet printing of highly loaded particulate suspensions. *Mrs. Bulletin, 28*(11), 815–818.

Forbes. (2015). Roundup of 3D printing market forecasts and estimates. Retrieved from http://www.forbes.com/sites/louiscolumbus/2015/03/31/2015-roundup-of-3d-printing-market-forecasts-and-estimates/#6e27547d1dc6.

Gibson, I., Rosen, D. W., & Stucker, B. (2010). *Additive manufacturing technologies.* New York: Springer.

Goodridge, R. D., Dalgarno, K. W., & Wood, D. J. (2006). Indirect selective laser sintering of an apatite-mullite glass-ceramic for potential use in bone replacement applications. In. *Proceedings of the Institution of Mechanical Engineers. Part H: Journal of Engineering in Medicine, 220*(1), 57–68.

Goodridge, R. D., Tuck, C. J., & Hague, R. J. M. (2012). Laser sintering of polyamides and other polymers. *Progress in Materials Science, 57*(2), 229–267.

Guo, N., & Leu, M. C. (2013). Additive manufacturing: Technology, applications and research needs. *Frontiers of Mechanical Engineering, 8*(3), 215–243.

Harrysson, O. L., Cansizoglu, O., Marcellin-Little, D. J., Cormier, D. R., & West, H. A. (2008). Direct metal fabrication of titanium implants with tailored materials and mechanical properties using electron beam melting technology. *Materials Science and Engineering: C, 28*(3), 366–373.

Hopkinson, N., & Dickens, P. (2001). Rapid prototyping for direct manufacture. *Rapid Prototyping Journal, 7*(4), 197–202.

Hopkinson, N., Hague, R., & Dickens, P. (Eds.). (2006). *Rapid manufacturing: an industrial revolution for the digital age.* Wiley.

Hull, C. W. (1990). U.S. Patent No. 4,929,402. Washington, DC: U.S. Patent and Trademark Office.

Khorram Niaki, M., & Nonino, F. (2017a). Additive manufacturing management: A review and future research agenda. *International Journal of Production Research, 55*(5), 1419–1439.

Khorram Niaki, M., & Nonino, F. (2017b). Impact of additive manufacturing on business competitiveness: A multiple case study. *Journal of Manufacturing Technology Management, 28*(1), 56–74.

Kruth, J. P., Wang, X., Laoui, T., & Froyen, L. (2003). Lasers and materials in selective laser sintering. *Assembly Automation, 23*(4), 357–371.

Levy, G. N., Schindel, R., & Kruth, J. P. (2003). Rapid manufacturing and rapid tooling with layer manufacturing (LM) technologies, state of the art and future perspectives. *CIRP Annals-Manufacturing Technology, 52*(2), 589–609.

Liu, J., & Li, L. (2004). In-time motion adjustment in laser cladding manufacturing process for improving dimensional accuracy and surface finish of the formed part. *Optics & Laser Technology, 36*(6), 477–483.

Liu, Q., Leu, M. C., & Schmitt, S. M. (2006). Rapid prototyping in dentistry: technology and application. *International Journal of Advanced Manufacturing Technology*, 29(3–4), 317–335.

Mansour, S., & Hague, R. (2003). Impact of rapid manufacturing on design for manufacture for injection molding. In. *Proceedings of the Institution of Mechanical Engineers. Part B: Journal of Engineering Manufacture, 217*(4), 453–461.

Monzón, M. D., Ortega, Z., Martínez, A., & Ortega, F. (2015). Standardization in additive manufacturing: activities carried out by international organizations and projects. *The International Journal of Advanced Manufacturing Technology, 76*(5–8), 1111–1121.

Mueller, B., & Kochan, D. (1999). Laminated object manufacturing for rapid tooling and patternmaking in foundry industry. *Computers in Industry, 39*(1), 47–53.

Mumtaz, K. A., & Hopkinson, N. (2010). Selective laser melting of thin wall parts using pulse shaping. *Journal of Materials Processing Technology, 210*(2), 279–287.

Murphy, E. J., Ansel, R. E., & Krajewski, J. J. (1989). U.S. Patent No. 4,844,144. Washington, DC: U.S. Patent and Trademark Office.

NASA/Jet Propulsion Laboratory. Accessed in 2017 from: https://www.jpl.nasa.gov/.

Nee, A. Y. C., Fuh, J. Y. H., & Miyazawa, T. (2001). On the improvement of the stereolithography (SL) process. *Journal of Materials Processing Technology, 113*(1), 262–268.

Petrovic, V., Vicente Haro Gonzalez, J., Jorda Ferrando, O., Delgado Gordillo, J., Ramon Blasco Puchades, J., & Portoles Grinan, L. (2011). Additive layered manufacturing: Sectors of industrial application shown through case studies. *International Journal of Production Research, 49*(4), 1061–1079.

Rosochowski, A., & Matuszak, A. (2000). Rapid tooling: The state of the art. *Journal of Materials Processing Technology, 106*(1), 191–198.

Shi, Y., Li, Z., Sun, H., Huang, S., & Zeng, F. (2004). Effect of the properties of the polymer materials on the quality of selective laser sintering parts. In. *Proceedings of the Institution of Mechanical Engineers. Part L: Journal of Materials Design and Applications, 218*(3), 247–252.

Singh, S., Ramakrishna, S., & Singh, R. (2017). Material issues in additive manufacturing: A review. *Journal of Manufacturing Processes*, 25, 185–200.

Wang, J., & Shaw, L. L. (2005). Rheological and extrusion behaviour of dental porcelain slurries for rapid prototyping applications. *Materials Science and Engineering: A, 397*(1), 314–321.

West, J., & Kuk, G. (2014). Proprietary benefits from open communities: How MakerBot leveraged thingiverse in 3D Printing. Available at SSRN 2544970.

Wohlers, T. (2007). *Wohlers report 2007: State of the industry: Annual worldwide progress report*. USA: Wohlers Associates Inc. ISBN 0-9754429-3-7.

Wohlers, T. (2014). *Wohlers report 2014: 3D Printing and additive manufacturing state of the industry*. Fort Collins, CO, USA: Wohlers Associates Inc.

Chapter 2
Industries and Applications

This chapter provides details of AM application sectors. It starts with the status of the technology in the global economic market and goes on to discuss three leading application sectors, namely the healthcare, automotive and aerospace industries as well as applications relating to consumer goods. Empirical evidence and the status of AM implementation in these industries is thoroughly described. It also outlines the main advantages to each industry of adopting AM technologies. The chapter discusses other application sectors as well, including building and architecture, food, and pharmaceutical industries along with applications for education and research institutes.

2.1 The Role of Additive Manufacturing in the Industry of the Future

The world has already seen three industrial revolutions in its history. The first dates back to the late eighteenth century and coincided with the introduction of the steam engine. The second is historically allocated to the last thirty years of the nineteenth century, with the advent of electricity that allowed the first mass production, and the third, since 1970, with the massive use of information and communication technology in plants. The so-called fourth industrial revolution, or Industry 4.0, is still underway. The term "Industry 4.0" was coined in Germany at the prestigious Hannover Industrial Fair in 2011. Since then, both national governments and private companies have launched numerous initiatives to develop this new manufacturing paradigm.

Policymakers, practitioners and scholars agree that Industry 4.0 is a new manufacturing paradigm triggered by new disruptive technologies. Baur and Wee (2015) reported for McKinsey on the four disruptions that are driving Industry 4.0: (1) the astonishing rise in data volumes, computational power, and connectivity,

M. Khorram Niaki and F. Nonino, *The Management of Additive Manufacturing*, Springer Series in Advanced Manufacturing, https://doi.org/10.1007/978-3-319-56309-1_2

(2) the emergence of analytics and business-intelligence capabilities, (3) new forms of human-machine interaction and (4) improvements in transferring digital instructions to the physical world. The fourth clearly relates to 3D printing and additive manufacturing technologies. In March 2015, the World Economic Forum's Global Agenda Council on the Future of Software and Society launched a survey aimed at identifying megatrends and technological tipping points—moments when specific technological shifts hit mainstream society. The report produced in September 2015 proposed six megatrends:

1. People and the internet
2. Computing, communications and storage
3. The internet of things
4. Artificial intelligence (AI) and big data
5. The sharing economy and distributed trust
6. The digitization of matter

The sixth megatrend refers to physical objects "printed" from raw materials via additive, or 3D printing. Moreover, the report proposes twenty-one tipping points in a resulting timeline ranging from 2018 to 2027 (Fig. 2.1).

Among the resulting overview of the expectations, the report contains the following tipping points in 2025 related to AM technologies:

- The first 3D-printed car in production (84.1% of agreement)
- 5% of consumer products printed in 3D (81.1%)
- The first transplant of a 3D-printed liver (76.4%)

During more than two decades of commercializing AM technology, the compound annual growth rate (CAGR) of worldwide revenues of all AM products and services was 25.4% in 2013. The growth rate increased by 27.4% over the three-year-period of 2010–2012, reaching \$2.2 billion in 2012. The number of AM systems sold (unit price > \$5000) increased by 19.3% to 7771 in 2012, while the unit sales of 3D personal printers (unit price ≤ \$5000) increased by 46.3% to

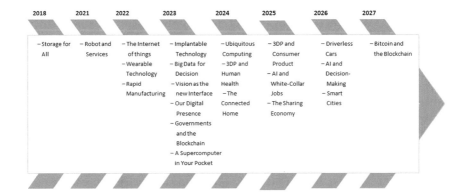

Fig. 2.1 Timeline of the twenty-one tipping points that will change industry and society. *Source* (Based on the survey conducted by World Economic Forum 2015)

35,508 in the same year. It demonstrates the spread of the technology for low-price 3DP among the retailers and end consumers. The AM market also continued its increasing growth in 2014. The market for AM, involving all AM products and services worldwide, grew at a CAGR of 35.2% to $4.1 billion in 2014, according to Wohlers and Caffrey (2015). The industry expanded by more than $1 billion in 2014, with forty-nine manufacturers producing and selling industrial-grade AM machines. The CAGR over the past three years (2012–2014) was 33.8%.

According to Wohlers' research, the global AM industry is expected to grow from $3.07 billion in revenue in 2013 to $12.8 billion by 2018, and exceed $21 billion in worldwide revenue by 2020. Siemens also predicted that 3D Printing would be a €7.7 billion ($8.3B) global market by 2023. Forecasting the market trends of future years has been difficult, considering the progressive growth during the recent period. Manyika et al. (2013) from the McKinsey Global Institute reported further adventurous predictions, estimating that AM could generate an economic impact of $230 billion to $550 billion per year by 2025, based on reducing costs and the value of customization; the largest source of potential impact would be from consumer uses.

SmarTech Publishing is one of the leading industry analysts in the AM environment, producing many reports per year pinpointing the major opportunities in the sector. The institution expects for 2017, that the dental and medical application sectors will drive a significant portion of the industry's market growth. In the healthcare industry, AM is becoming a viable and accepted technology, with a growing presence particularly in the surgical sphere (e.g., cardiac and orthopedic uses).

2.2 Industrial Diffusion

Researchers are currently demonstrating the rapid spread of AM technologies, and according to the results of the survey of more than 100 industrial manufacturers made by PwC (PricewaterhouseCoopers) Technology Forecast, approximately 67% of manufacturers are already using 3D printing (Fig. 2.2). Of these, 28.9% are experimenting to determine how 3D printing can be integrated into their current production, and 24.6% are using 3D printing for prototyping.

Another piece of research investigated the reasons for pursuing AM technologies. In 2014, Gartner, Inc. conducted a worldwide survey, involving 330 individuals. The participants were employees of organizations with at least 100 employees that are using or planning to use AM technologies (Fig. 2.3). The results show more than half were already using AM in the initial stage of manufacturing, aimed at the product development process. An interesting finding of this research is that 60% of the respondents said that high start-up costs are a main reason for delays in implementing the technology. However, most of the early adopters found clear benefits in different aspects while adopting AM technologies. It shows that more and more organizations will be attracted to use 3D printing as technical

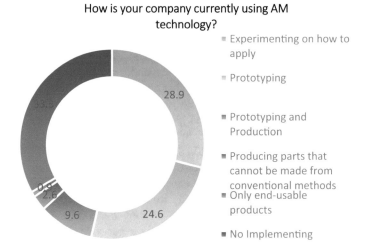

Fig. 2.2 A survey on the AM industrial diffusion level, 2014 (Based on the data of PwC)

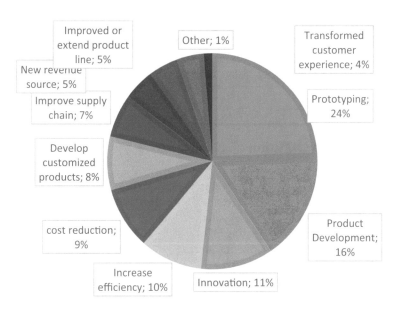

Fig. 2.3 Reasons for pursuing AM technology, 2014 (Based on the data from Gartner, Inc)

improvements are made and start-up costs reduce, According to the PwC estimation, 67% of the manufacturers already use AM technology, and given technological improvement in recent years, we can estimate a greater adoption rate among
manufacturers in the near future. Considering all the target communities, involving
consumers, industries and life science, Gartner also predicts almost 50% of them

will adopt AM technology—by the year 2018—in their services and in the production of parts for the items they consume and sell.

Furthermore, a SmarTech market publication in May 2014 investigated and predicted the AM market in detail. The total AM market is divided into four sections, constituting software, services, material, and equipment, which respectively contributed 5, 23, 26, and 46% of the AM market share in 2013. In addition, it predicts that by 2023, the AM market share for each will be respectively 10, 14, 28, and 48%. It shows also the necessity of focusing on software and material development to produce a more powerful and more widespread manufacturing technology. Another interesting analysis was performed in order to distinguish the industrial sectors inside AM markets. For instance, the automotive industry's adoption of 3D printing is projected to increase from $365.4 million in 2015 to $1.8 billion in 2023, making a 19.51% CAGR. The aerospace industry's adoption of 3D printing solutions is projected to increase from $723 million in 2015 to $3.45 billion in 2023, a CAGR of 18.97%.

Wohlers (2012) reports the share of each application sector (Fig. 2.4). Automotive, aerospace, medical, consumer products, and other industrial parts have respective shares of 17.3, 12.3, 13.7, 18.5, and 18%. The detailed discussion of these industrial sectors will be presented in Sect. 2.3.

Gartner et al. (2015) conducted a survey of 409 AM adopters among different industries and application sectors. The respondents were asked to answer regarding the level of impact of AM technology on the industry. Figure 2.5 illustrates the list of application sectors and their answers ranging from "AM is expected to have a strong hindering effect," to 'AM is expected to have a strong fostering effect," (on your industry) within the next ten years. The results demonstrate that some application sectors are more optimistic about the future of AM rather than others. These application sectors, of which more than 80% perceived the fostering effects of AM, include consumer goods (i.e., hobbies and models, arts and fashions), the automotive, aerospace, and healthcare industries among others, and research institutions.

Fig. 2.4 AM industrial market share

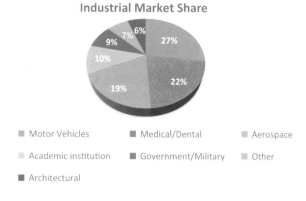

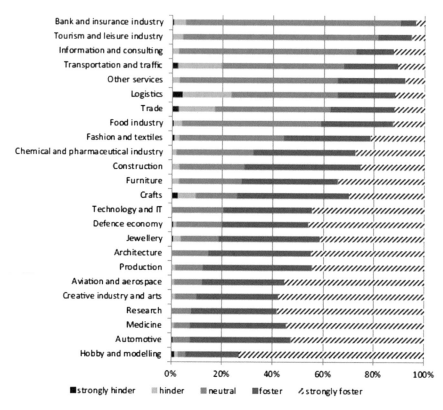

Fig. 2.5 Ten-year forecast of the impact of AM by Industry (*n* = 407). *Source* Gartner et al. (2015)

2.3   Main Application Sectors

2.3.1   Healthcare Industry

Medical products are one of the leading AM application sectors because medical products usually require a customized shape and functionality. As reported, nearly 15.1% of the U.S. AM market takes place within this industry, with AM shipments estimated to be $37.2 million, less than 0.05% of medical and dental manufacturing, and it also includes 16.4% of worldwide AM markets in 2013 (Wohlers 2014). AM has also been employed worldwide in the creation of around 30,000 prosthetic limbs, more than 500,000 dental implants, and numerous other devices (Campbell et al. 2012).

Since AM allows for fast and economic production of customized designs, it is well matched to applications in this sector. For instance, the ability to produce custom implants eliminates the need for time-consuming adjustments during surgery and reduces operating costs as well as the risk of medical complications

(Bogue 2013). In addition, the flexibility provided by AM allows surgeons to decide on the most proper part for each patient, and to simply modify and optimize the design, resulting in the improved performance of implants.

Giannatsis and Dedoussis (2009) classified the applications of AM in the medical sector as follows:

- Bio modeling, involving the fabrication of physical models of parts of the human anatomy and biological structures in general, for surgery planning or testing
- Design and fabrication of customized implants for prosthetic operations, rehabilitation, and plastic surgery
- Fabrication of porous implants (scaffolds) and tissue engineering
- Fabrication of specific surgical aids and tools
- Drug delivery and micron-scale medical devices

The use of AM in bio modeling for the purpose of surgery planning or for diagnostic uses was probably the first medical AM application (Gibson et al. 2010). Fabricating a model, which allows surgeons to see from any angle and can touch is very valuable. They use these custom-made models to identify complex surgical procedures. These models are usually first imaged by 3D imaging machines such as CT (computerized tomography—used to generate cross-sectional images, of the bones, blood vessels and soft tissues inside the body) or MRI (Magnetic Resonance Imaging, which uses radio waves to create an image of the organs), and are then transferred to the AM machine. Surgeons stated that having a multicolored, complex model of the head or abdominal area were priceless in planning surgeries, a process which can take 12–24 h and involve teams of surgeons and support staff. Digital personalized surgery (Fig. 2.6) is a workflow guiding the whole process from the medical imaging of the patient through to the surgery, transitioning from the virtual world to the physical world using perceptual design tools and software, surgical simulation and 3D printed surgical guides and models.

AM has found applications in a variety of surgeries such as pelvic surgery, neurosurgery, spinal surgery, cardiovascular surgery, and visceral surgery, while research demonstrates a significant improvement in diagnosis and treatment thanks to better 3D models of pathological structure, increased accuracy and the possibility

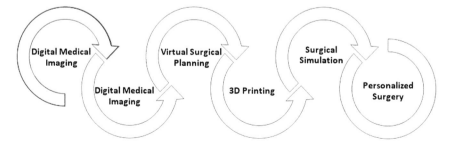

Fig. 2.6 Workflow of digital and personalized surgical planning

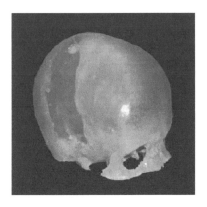

of pre-planning (Rengier et al. 2010). An example of a skull model, made by SLA, is presented in Fig. 2.7. This biomodel is an accurate replicate of the damaged skull of a young girl, which was used for pre-operative planning of the surgery and an aid for the design of the prosthetic implant. The first application phase included the use of AM for rapid tooling (RT) purposes. Then through the improvement of the technology and corresponding quality matters, AM technologies are employed in a variety of end-use parts rather than in models.

Manufacturers offer various devices and components for electrotherapy. For instance, one of the companies (Ivivi Technologies) uses AM technology in its product development process, yielding a positive return on investment in less than one year. The company argues that AM empowers designers to make product changes overnight. Therefore, AM is an amazing manufacturing method, in which manufacturers can modify products to meet their customers' needs in only a few hours, resulting in a boost to their customer relationships. In addition, AM gives the company a rapid and cost effective way of manufacturing customized medical parts, mostly by eliminating the need for tools.

AM is also well suited to the production of prosthetics due to its ability to produce low volumes, complex designs, and customized parts. Although standard-sized implants are commercially available, these might not be usable in all cases. Rengier et al. (2010) argued the need for customized implants as follows: "(1) patients outside the standard range with respect to implant size- or disease-specific special requirements, (2) improved surgical outcome because of individual fitting and adequate match with individual anatomical needs". The series production of end-use parts has been applied in the fabrication of acetabular cups used as a hip joint replacement, produced using the Arcam EBM system. The use of AM allows the incorporation of embedded porosity to promote cell ingress.

AM has been also used in series production of dental parts. Vandenbroucke and Kruth (2007) state that "Dental applications are very suitable to be produced by SLM due to their complex geometry, strong individualization and high-aggregate price. Moreover, the manufacturing of multiple unique parts in a single production run could enable mass customization". 3D printing is finding growing applications in producing customized crowns, implants and bridges. A successful application of

RM is in orthodontic treatment, which involves the patient wearing a sequential series of customized clear plastic aligners. Each aligner in the series is produced indirectly from individual SL models. The adoption of AM in dental industries has an increasing number of advantages. For instance, NimroDENTAL Orthodontic Solutions is a service supplier to dentists in the London area. The company produced accurate functional models for aligners, crowns, and bridges more rapidly than with any previous method, using transparent and colored AM raw materials. In addition, the higher contrast of parts made using available materials compared to the traditional plaster models, results in better customer relationships (see Fig. 2.8). These models enable the patients to easily see how the orthodontic device looks and works. According to the experience of the company, better communication with patients as a result of the use of 3D printed models had a real and positive impact on customer satisfaction. They can 3D print up to twenty high-quality dental models on demand at short notice, saving time, lowering costs and reducing the need to store large numbers of physical models in the clinic. Furthermore, considering the nature of clinical areas, the size of 3DP and the cleanliness of production make it more effective than plaster model making.

AM has also found an application in the hearing aid industry (Fig. 2.9). For instance, Siemens and Phonak are both using AM systems for the series production of customized hearing aids. Two systems are currently available for the fabrication of customized hearing aids, SLS and SL (Hopkinson et al. 2006). AM technology is also being used routinely for the manufacture of plastic hearing aid parts, ranging from individual ear molds to shells with integrated faceplates. The success of AM in the hearing aid industry has been a particularly impressive example of how companies can take advantage of the shape complexity capability of AM technologies to

Fig. 2.8 Sample dental molds made by 3DP

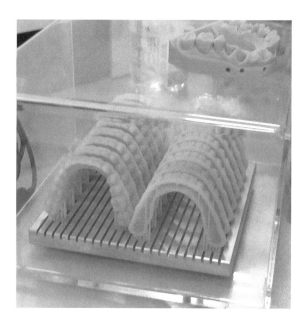

economically achieve mass customization (Gibson et al. 2010). The AM replacement offers an almost fully automated solution, with a lead time down to a day and a "first-fit" rate of 95% (Ruffo and Hague 2007). According to Dr. Phil Reeves, reported in *3D Printing Industry* in 2013, "there are more than 10,000,000 3D printed hearing aids in use worldwide". Jenna Franklin, marketing associate of EnvisionTEC, a leading manufacturer of 3D printers, claims that the majority of existing hearing aids in the world are fabricated using 3D printers.

In recent years, scientists have developed a means of using a patient's cells to enable bioprinting of skin tissues and a range of other human body parts. Bioprinting involves placing living cells into just the right location with an AM machine, to fabricate functional, heterogeneous living tissue. In February 2013, experts from Weill Cornell Medical College used AM and injectable gels made of live cells to build a facsimile of a human ear (PrototypeToday 2013). According to a medical expert, "the ability to manufacture living human tissue for medical research and clinical practice has the potential to reshape the future of medicine" (IndustryWeek 2013). One of the major advantages of using AM in bioprinting with injectable gels, is a printer's ability to carefully squirt cells into precise patterns and shapes. In addition, the ability of AM machines to simultaneously print in multi-materials, means that biological researchers are a step closer to creating artificial tissue that mimics nature's complicated shapes, internal structures, and cellular diversity (Lipson and Kurman 2013).

Furthermore, in recent years a range of materials has been introduced. There are three types of polymers, used in the medical sector. These are biocompatible polymers for external use, biocompatible polymers for implants, and biodegradable polymers, also for implants (Petrovic et al. 2011). Biocompatible polymers for

Fig. 2.9 A hearing aid made by 3DP. *Image Source* Gibson et al. (2010)

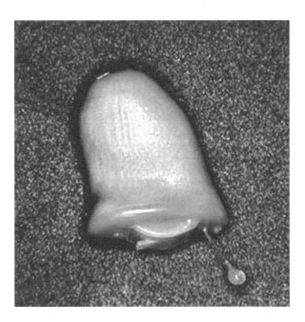

external use involve those that can be used for medical or prosthetic devices. For instance, the e-shell 200 series material has been developed particularly for use in hearing aid products, because of its rigidity and durability. Other biocompatible polymers involve those that can be used in surgical implants. These materials do not result in rejection by the human body. An example of this group is the photopolymer R 11, used in the fabrication of coronary bypass implants. Hard tissue lumbar implants made with SLA for vertebral fixation have also been implanted (Goffin et al. 2001). In addition, several research studies have been conducted in order to obtain scaffolds for tissue engineering using polylactic glycolic acid (PLGA) and fibrin gel (Liu et al. 2009), or a combination of ether, epoxy resin and a photo initiator (Quadrani et al. 2005). Biodegradable polymers can be used in implants with the special feature of an osteoinductive function that allows bone growth and degrades over time. In addition, the flexible form of these polymers permits tissue engineering, as they can be used as a basic material for fabricating human organs (Petrovic et al. 2011). On the other hand, metal implants interface with living bone and tissue, with EBM and DMLS technologies both now used in the production of standard and customized implants. For instance, metals like titanium are used particularly in load-bearing areas, for instance in hip reconstruction (Rengier et al. 2010). Biocompatible materials include bioceramics such as hydroxyapatite which is currently the preferred material for bone reconstruction (Stevens et al. 2008). Commonly used biomedical materials for use in medical components are listed in Table 2.1. Moreover, several AM systems have been adopted in medicine including EBM, FDM, SLA, SLS, and DMLS.

In general, the emergence of AM in the medical sector provides a number of advantages. One of the most important of these is that the need to fabricate complex or customized designs in either single units or small volumes is already provided by AM. In addition, the product development process will be faster than with any conventional manufacturing method. When using AM in the medical sector, we usually start with a CT scan of the required zone, which is then transferred to an AM machine to be prepared. In addition, AM is able to fabricate gradual structures enabling production of lightweight parts, along with osteoinduction and osteointegration in implant manufacturing.

2.3.2 Automotive Industry

AM technologies have been employed in the automotive industry for a decade in prototyping, tooling and the production of special and customized parts in small volumes. AM shipments for the U.S. automotive industry were worth $48 billion in 2011. As mentioned above, approximately 19.5% of worldwide AM occurred within the automotive industry in 2012, with AM shipments estimated to be less than 0.05% of total U.S. automotive shipments (Wohlers 2012).

Due to the significant advantage provided by AM in the production of customized and small volume parts, RM applications were well established in

Table 2.1 Available biomedical materials and applications

Implant material	Abbreviation/notation	Application
Ti and Ti alloys	CP–Ti	Bone fixation
	Ti–6A1–4 V	Artificial valve, stent, bone fixation
	Ti–6Al–7Nb	Dental application, knee joint, hip joint
	Ti–5Al–2.5Fe	Spinal implant
	Ti–15 Zr–4Nb–2Ta–0.2Pd	Dental applications
	Ti–29Nb–13Ta–4.6Zr	Dental applications
	83%–87%Ti–13%–17%Zr (Roxolid)	Dental applications
Stainless steel	316L	Dental, knee joint, hip joint, surgical tools
Cobalt chromium alloy	Co–Cr–Mo, Co–Ni–Cr–Mo	Artificial valve, bone fixation, dental applications, knee joint, hip joint
Shape memory alloy	NiTi	Catheters, stents
Polymers	PMMA, PE, PEEK	Dental applications, articular cartilage, hip joint bearing surface, knee joint bearing
Bio-glass	$SiO_2/CaO/Na_2O/P_2O_5$	Dental applications, orthopedic implants
Zirconia	Zirconia	Porous implants, dental applications
Alumina	Al_2O_3	Dental applications
Hydroxyapatite	$Ca_5(PO_4)_3(OH)$	Dental applications, implant coating material

Source Singh et al. (2017)

motorsport rather than in conventional automotive manufacturing. However, AM found widespread application in automotive industries similar to that in other industrial sectors for prototyping and tooling (RP/T). Using AM removes design constraints currently imposed on the automotive designer by tooling limitations (khorram Niaki and Nonino 2017). The ability to fabricate fully individualized parts has a particular impact on ergonomic design, which is based on the customer's specific requirements and comfort fit. One example here is the use of AM in Formula 1 (F1) car racing. The Renault F1 team was the first team that established a manufacturing center that uses AM. They produce a number of car parts directly with AM technology, with beneficial effects of reducing the need for assembly. For instance, brake cooling ducts were produced by AM as the geometrical complexity makes it impossible for them to be produced by conventional manufacturing methods in one piece (Kochan 2003). Figure 2.10 shows a cooling duct, the new design replaces 16 parts with 1 part.

Kingston University's electric car racing team (KU e-Racing) is emerging as a leader in the field. In 2013 and 2014, it was named the United Kingdom's highest scoring Formula Student electric team in the annual Formula Student race at Silverstone. Using AM, they were able to significantly reduce the overall weight of the car by replacing metal parts with lightweight printed plastic components. They

(a)   **(b)**

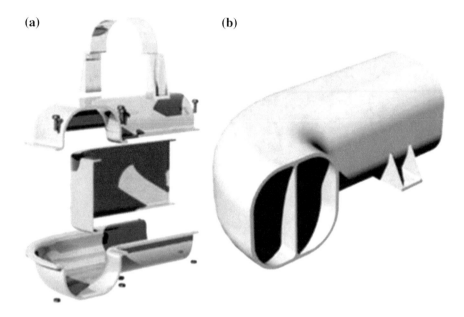

Fig. 2.10 A typical cooling duct: **a** Original design with 16 parts, **b** Integrated design. *Image Source* Gibson et al. (2010)

also 3D printed parts with the objective of fitting them directly to the car, thanks to the precision of 3DP. In the process, AM enabled the company to reduce the cost through the simultaneous manufacture of a number of production parts. It resulted in extreme reduction in labor costs and production time of around 40% compared to a conventional manufacturing process.

Thanks to AM, the company was also able to design and fabricate more complex parts than designs allowed when using conventional manufacturing because AM enables more freedom of design and increased creativity without the limitations of conventional manufacturing (see Sect. 5.1.3 for more detail). A design engineer from the team said, "With 3D printing, we could design parts that adapted to our specific needs rather than designing the car around the parts". Ku e-Racing 3D printed an exhaust fan housing in high-performance ULTEM 9085 resin, for automotive applications with its FST (flame, smoke and toxicity) rating and high strength-to-weight ratio. Moreover, the company efficiently 3D printed a variety of racing car components such as a shutdown button mounting in ABS plus thermoplastic, and an air intake system in PC-ABS. Figure 2.11 shows a complex race car upright that was fabricated by EBM in just 28 hours, demonstrating the advantage of AM for reducing time to market and increasing the flexibility.

For luxury cars with low production volumes, AM can also be economical in the manufacture of custom parts or replacement parts for antique cars. For instance, Bentley Motors produced some customer interior components, such as bezels in this fashion. Typically, Bentley production volumes for the given model are lower than

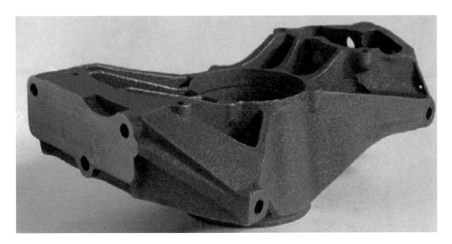

Fig. 2.11 A race car upright made by EBM. *Image Source* Petrovic et al. (2011)

10,000 cars so this could be considered as a medium production volume. Lamborghini S.p.a., a leading manufacturer of super sports cars, uses AM technologies throughout the entire lifecycle of its parts, from rapid prototyping to rapid manufacturing applications. They purchased the first 3DP (a Dimension 1200es from Stratasys) in 2003. The company mainly employed AM with the objective of rapidly fabricating parts that have complex geometries, as well as to meet the demand for high-strength production parts tough enough to endure the rigors of high-speed racing. As stated by the technical department of the company, 3D printed parts (i.e. profiles and air channel) meet the required quality level for high-performance aesthetic parts. AM also has accelerated the RP process by reducing costs and enhancing workflow efficiency. It also provides the company with a greater level of control over projects compared to the previous classic route of outsourcing the prototyping and model making. In addition, Lamborghini applied SLS of carbon fiber filled material to produce headlight washer cover flaps (Fig. 2.12) for pre-production models of the Lamborghini Gallardo for immediate delivery to dealers and customers. In these cases, AM provides a shortening of the lead time for low volume series production.

On the other hand, since the production volume in the automotive industry is usually high (100,000 per year) and commercial road vehicles have high quality requirements, AM applications are therefore being confined to RP and RT. AM has been evaluated as an expensive manufacturing technique for mass production, however, for product development processes, it brings several competitive advantages. AM enables a reduction in the time to market in a predictable manner, resulting in saved time and costs in the new product development (NPD) process. An example of this approach is the use of AM as part of a bridge-tooling process to make parts available before the full production tool is ready. Campbell et al. (2012) stated that it is obvious that AM is now a core technology for automotive product development. One of the early adopters within the industry was the Ford Motor

Fig. 2.12 Lamborghini headlight washer cover flap. *Image Source* CPR technologies

Company, which has used AM in the prototyping and testing of vehicles since the 1980s. They produced prototypes of cylinder heads, brake rotors and rear axles in a reduced time in comparison with conventional manufacturing methods and they succeeding in saving about one month of production time when creating a casting for a prototype of a complex cylinder engine head that included multiple ports, ducts, passages, and valves to manage fuel and air flow (Boulton 2013). In addition to these applications, suppliers to the automotive industry use AM parts to debug their assembly lines, and for testing their operations and tooling in order to identify potential problems before starting production. Since model line changeover involves huge investments, being able to avoid problems in production yielded very large savings (Gibson et al. 2010).

The above-mentioned areas of application include the use of AM in prototyping, tooling and small-volume high-value part production, and the advantages of each area were discussed. It is mentioned that using AM for mass production is seen as an expensive manufacturing method. However, the technology is dramatically developing in terms of size, speed and range of materials, and the growing investments from large companies in AM also demonstrate the increasing potential of this industry.

For example, Daimler AG recently funded the development of the X Line 1000R System of Concept Laser GmbH, a powder-based laser-fusing machine with a large size, for constructing metal components for vehicles and engines. The machine has a build chamber of 23.6 *15.7 * 19.7 in. (large enough to accommodate a V6 auto-motive engine block) and a layer thickness of 20–100 microns. In addition, the production rate of the 1000R is 65 cm^3 per hour, which is faster than other powder-based fusion 3D printers. The new machine involves a high-powered laser, allowing as much as ten times the productive capacity of other machines. Further understanding of the capability of AM came from KOR Ecologic Inc., which intro-duced a two-seater "green" car named the "URBEE 2" (Fig. 2.13). Its entire exterior and interior are 3D printed. "A future where 3D printers build cars may not be far off after all," said Jim Bartel, VP. An initial prototype was completed in 2013 and it became the first car to have its body 3D printed, and is the "greenest" car ever made.

Fig. 2.13 3D printed
URBEE2 car body. *Image
Source* KOR Ecologic Inc.

2.3.3 Aerospace

The aerospace industry, like the automotive industry was among the early adopters
of AM. They adopted the technology mainly for RP from 1990s. AM has significant potential in the aerospace industry, which needs parts with complex geometries, low weights, and strength. Typically, aerospace operations involve high-value
and low-volume production that is especially suited to the main benefits offered by
AM. In 2012, approximately 12.1% of AM occurred in the aerospace industry. AM
shipments for the U.S. aerospace industry were $29.8 million, or less than 0.05% of
the U.S. aerospace industry (Ford 2014). According to Ford (2014), adopting AM
technologies led to a wide variety of impacts on the aerospace industry such as a
reduction in material usage, a simplified spare part supply chain, and the enabling of
extremely complex geometries and integrated parts.

Reduced material usage
    The idea of fabricating parts with a lattice structure in their interior to reduce
weight is an issue in several industrial sectors. The strength-to-weight ratio is particularly significant in aerospace industries, where an additional 100 kg of weight,
can increase the expense of an airline by more than $2.5 million in fuel over the
aircraft's lifetime (Hopkinson et al. 2006). Usually, producing these structures with
conventional manufacturing is difficult—or in some cases impossible. However, AM
enables the direct fabrication of lattice structures with gradual and controlled porosity
(Petrovic et al. 2011). Moreover, aerospace parts are very expensive and as a result
they are developing a significant proportion of their parts from titanium, plastic, and
other lightweight materials. The conventional processes are wasteful, sometimes
exceeding 90%, while the material usage efficiency of AM is reported to be 97%
(Achillas et al. 2015). According to Campbell et al. (2011) aircraft landing gear has
been additively manufactured from titanium rather than cut from a titanium block,
resulting in reduced material waste and costs. The buy-to-fly ratio is another indicator

in this industry, which means how much material is needed to be purchased in order to manufacture the final part. The buy-to-fly rate in this industry reaches 15–20, in other words 15–20 kg are required to produce 1 kg of end product, while using AM reduces this rate to nearly 1 (Cozmei and Caloian 2012). This measure indicates that AM possesses a basic characteristic of economical operation. In another study, Reeves (2008) demonstrated that the same part, redesigned for SLM, had the same mechanical properties with 40% less material. Consequently, the advantages of AM in these cases involve reducing the use of raw materials, reducing the buy-to-fly ratio, and reducing corresponding energy consumption.

Spare part supply chain

The demand of aerospace spare parts normally follows a 20/80 Pareto curve (Liu et al, 2014). This means that 80% of the parts which are needed frequently, make most of the airline's demands. However, they only account for 20% of the supply chain expenditure in terms of holding inventory and moving materials. The supply chain of some aerospace spare parts is very slow and too unpredictable, resulting in a high cost. Therefore, many aerospace companies have applied AM technologies for these types of parts, which are high-cost long-lead components, by producing parts on-demand. Furthermore, many aerospace parts have a long useful lifecycle, therefore AM enables the use of digital models of parts instead of having to store that part itself during this long period. The impact of AM on the reduction of supply chain safety-inventory was analyzed quantitatively and the potential of AM to make the aircraft supply chain more efficient was clearly demonstrated (Walter et al. 2004; Hasan and Rennie 2008; Holmström et al. 2010). The detail of supply chain approaches, considered in light of the emergence of AM, will be discussed in Chap. 5.

Assembly requirements

Integrated parts are an ideal for the designer. However, it is impossible in some cases due to conventional design and molding constraints. Because of the freeform nature of fabrication in AM, it presents a new opportunity in optimized design for manufacturers. One industrial example of this impact has been reported by Campbell et al. (2011) who detailed how the environmental control system duct on the F-18 has been redesigned taking into account the capabilities of AM and the number of parts has been reduced from sixteen to just one. Although the conventionally manufactured assembly must have its design tailored to fit the capabilities of the machine tools used to produce the part, the AM part is built precisely to realize its function.

Complex geometry

Aeronautical components may also have more than one function. For instance, a turbine blade also has an internal structure for passing coolant through it. In addition, geometric specifications for parts may be determined by complex mathematical formulae based on fluid flow, etc. (Gibson et al. 2010). Currently, by using AM to fabricate the parts, dealing with these complexities is becoming more feasible.

A majority of companies in this sector have been using AM for many years, implementing this technology with different objectives. Rochus et al. (2007) classified these applications of AM as follows: Basic prototyping—scale models; mock-ups—geometric representations; test articles—for testing and qualification; and real flight hardware.

For instance, Boeing has used AM for the manufacture of more than 200 different components of ten aircraft platforms (Ford 2014). It has also installed over 20,000 AM parts in military and commercial aircraft (Wohlers 2014). Boing implemented thermoplastic SLS components for its 737, 747 and 777 commercial aircraft, in addition to thirty-two different components for its 787 aircraft (Freedman 2011). Boeing might only produce a small number of them during a year, therefore these were not items which have to be mass produced. Though the speed limitations of production with 3-D printing might keep it from ever producing the majority of Boeing's parts, Vander Wel says that the approach is likely to be used in a growing proportion of them. Moreover, as for metal parts, General Electric (GE Aviation), which is the world-leading provider of jet engines, invested in 2012 with the aim of developing metal AM parts for use in a gas turbine engine. GE produced a fuel nozzle for use in thousands of jet engines with AM technology (Ford 2014). The redesigned nozzle is 25% lighter and has a five times more useful lifetime than existing models, which are welded from twenty different parts (Gibson et al. 2010). In addition, the new design was engineered to reduce carbon build up, making the nozzle more efficient. The new design benefited from using less material leading to a reduction in weight, no assembly requirement and better quality. Additionally, this fuel nozzle is projected to reach a production volume of more than 100,000, where each engine contains nineteen nozzles, and GE has sold more than 4500 engines. Furthermore, GE reported that up to half of the parts in its energy and jet engines will be manufactured by AM within ten years. Figure 2.14 illustrates examples of 3D printed parts for aerospace industry including helicopters and jet engines.

Another leading aerospace company, Airbus, has additively manufactured the A320 nacelle hinge bracket, which was originally designed for steel casting. It reported that using AM for the redesigned part resulted in a 40% saving in material usage (Gibson et al. 2010). Generally, AM parts are beginning to appear on a range of Airbus aircraft, from the next-generation A350 XWB to in-service jetliners from the cornerstone A300/A310 family. Aurora Flight Sciences also fabricated and flew a 62-inch wingspan aircraft with a wing composed entirely of AM parts. Aurora expects to fabricate small, unmanned aerial vehicles, both military and civilian, within five years (Ford 2014).

Sheppard Air Force, the Trainer Development Flight (TDF) manufactures training aids that are utilized for the Air Force. The training aid can be either an original product or a replica of an existing one. Often, using these replicas is more economical for training applications, rather than using the more expensive original ones. Traditionally, TDF has used a variety of conventional manufacturing processes, such as machining, sheet metal bending and cutting, welding and lathe work, to make its products. As the production volumes were very low, the conventional method became very expensive, similar to other single or low-volume

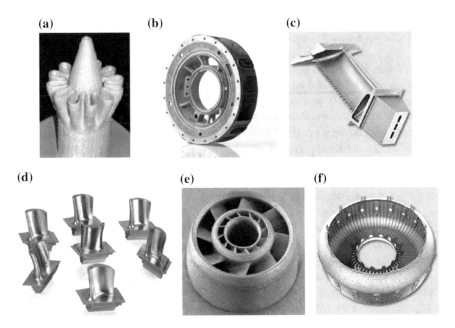

Fig. 2.14 Examples of 3D printed components for aerospace industry: **a** Mixing nozzle for gas turbine exhaust; **b** compressor support case for gas turbine engine produced by EBM; **c** turbine blade with internal cooling channels produced by SLM; **d** turbine blades fabricated by SLM; **e** hollow static turbine blade cast using the mold and cores fabricated by 3DP; **f** engine housing produced by SLM. *Image Source* Guo and Leu (2013)

production environments. Using FDM, one of the AM technologies, they could effectively manufacture many components of a replica of an unmanned aerial vehicle (UAV) such as most of the body components, several cowlings, propellers, and antennas.

According to the chief of TDF, the parts printed with FDM were durable and met their quality requirements. Moreover, it terms of environmental impacts, the process is safe and completely green producing zero waste. In terms of the profitability of investment, TDF produced the UAV's large antenna in one-tenth of the time taken by conventional manufacturing, and it delivered a return on investment (ROI) of over $12,000. Reportedly, conventional manufacturing took up to twenty days in an outsourcing process, while using FDM it takes only two days with fifteen to twenty minutes of the workforce. Generally, the impact of implementing AM in TDF was very impressive and resulted in a saving of $800,000 on the cost of projects over the four years. The CEO says, "For our first FDM machine purchase, we projected a ROI in four years, but it took only eighteen months", Weatherly says. "For our second FDM machine purchase, we saw a ROI in only nine months". In addition, TDF benefits from some other main advantages of AM, including the ability to run multiple parts simultaneously, using multiple materials for a variety of purposes, ease of maintenance, and the ability to modify designs at minimal cost.

Moreover, aside from the commercial aircraft applications, AM is an interesting manufacturing method for the space sector. In addition to all benefits mentioned above, the ability to shift the production location might be an important factor in using AM in space. It is critical when space explorers venture far from Earth and will create an on-demand supply chain for any tools and parts needed.

The National Air and Space Administration (NASA) considers AM as an essential tool for its space exploration plans, with 3D printers replicating themselves and making spare parts as well as large structures in space. Made-In-Space, a Silicon Valley start-up, prepared an AM machine that passed all NASA tests for certification. Its first zero-gravity 3D printer was delivered to the International Space Station (ISS) in 2014, and it is projected to return to Earth for further tests to verify that the 3DP works the same in microgravity as it does on Earth. Figure 2.15 shows the 3D Printer during testing in the Microgravity Science Glovebox (MSG) Engineering Unit at the Marshall Space Flight Center. NASA has also commissioned the development of 3D printers for food and for building structures on the Moon.

In addition, NASA and Pratt & Whitney Rocketdyne jointly additively manufactured a rocket engine injector, which is one of the most important and expensive components of a launch vehicle. They succeed in producing it by AM in four months and reduced costs by 70%, while conventional manufacturing took typically one year.

Fig. 2.15 Made in space 3D printer testing. *Image Source* Wikimedia

2.3.4   Consumer Goods

Theoretically, every object that is designed by computer-aided design software (CAD) can be converted to the final physical part with AM technologies using available materials. Consequently, it assists creativeness and makes the world more innovative. According to Wohlers (2012), consumer products and electronics constitute approximately 20.3% of additive manufacturing markets. This application sector will continue to growing given the development of affordable 3D printers and technological advancement. However, the prices of the machines and materials has decreased in recent years, and this is outlined in Chap. 1. For instance, the "Cubify Cube" provides a desktop 3D printer for about $1300. Nevertheless, the machine is unable to produce sturdy parts. The "MakerBot Replicator 2" is another option for home-use 3D printing. The machine is able to print in two different colors and uses materials including PLA, ABS and plastic. Nevertheless, there is an apparent need to develop better and cheaper 3D printing machines for consumer applications.

The lack of a medium in which designers would be able to directly convert their own design to the final products has existed for many years. It can be presumed that huge individual design ideas have been missed due to the lack of manufacturing facilities. AM addresses this issue and offers designers the enhanced possibility of transforming their designs into products in less time and without the need for conventional manufacturing. It also facilitates co-creation between customers and producers, and provides people with an opportunity to become micro-entrepreneurs. Simpler steps now need to be taken in order to reach the required objects, thanks to AM technology and 3D printing platforms (see Sect. 4.1.3 for more detail). Individuals can initiate a project by creating a CAD file that can be sold through AM marketplaces, such as Shapeways. Otherwise, they can modify existing CAD files that are available on websites (such as thingiverse.com) and sell 3D printed parts of these files. This will turn customers into what is called "accidental entrepreneurs" (Shah and Tripsas 2007).

AM found its own business community, the so-called Makers, which includes small business units who sell and share 3D models, 3D printed parts, and even collaborate with customers in design and production. Likewise, at the beginning of the emergence of the internet, there were many markets or internet centers facilitating the use of the internet for their customers. There are many start-ups, like Shapways, Makerbot, Materialise, etc. in this growing marketplace and they provide a variety of services (software and hardware) to their customers, from the initial ideas to the final products. These services include assisting in file preparation, optimizing design, selecting the proper technology and material, production process, and cloud-based design and production. GE Garage is another 3DP service supplier project, which began in 2012 with the objective to "reinvigorate America's interest in invention, innovation, and manufacturing" (GE 2012). GE Garage provides a platform to collaborate with designers and customers of AM. Stratasys, a leading 3D printing supplier, acquired some competitors to enter the consumer products' market. After the notable acquisition of Makerbot in 2012, Stratasys acquired Concepts Inc. and Harvest Technologies in 2014 to consolidate its portfolio in the consumer market. Inexpensive consumer 3D printers and open-source 3D designs might even change the way children play. Children can download 3D design files for the toys they want, modify and customize them as they wish, and then 3D print them in their homes.

In addition to the opportunities for self-fabrication, there are also many industrial manufactures, adopting AM for consumer goods. They may adopt whether alongside their conventional manufacturing methods or as a single manufacturing method. The following paragraphs provide examples for manufacturing of various consumer goods. 3T RPD Ltd., an AM service provider in England, fabricates its fully personalized gifts—a titanium tie clip of their customer's signature. The product's quality shows how the process deals with both angle and up-skin surfaces. Then, the very top surface of the signature is hand polished to highlight the text. This product demonstrates the individualization and customization that AM enables, likewise, another company offers the full customized bow ties (see Fig. 2.16).

Eye Venture Ltd, in collaboration with 3T RPD Ltd. developed a new product, called the ZYclip for both prescription eyewear and sunglass. Conventional manufacturing methods for the titanium-based frames included tooling, which was time consuming and expensive. Additionally, the machining process of the parts would

Fig. 2.16 Customizable bow ties

result in 95% of the material being wasted. Therefore, the company identified AM as an alternative method of manufacturing that enables them to eliminate the need for tooling and increases the flexibility in future interchangeable designs of the frames. It also was beneficial in design modifications that could be made to the CAD data, and the parts then rebuilt quickly and cost effectively. These have given Eye Venture the maximum flexibility to design eyewear frames to almost any shape or size and offer a real cost benefit over conventional manufacturing methods.

Companies are also implementing AM for the design process of this type of products. The Design Centre at Logitech Ireland Services was seeking to improve their design process and their first AM equipment was installed in 2003. The primary objectives involved testing the fit, form and function of the designed part. It enabled the design team to cost effectively explore their creative ideas and reduce the cost of design errors. Logitech employs AM throughout the design process. In concept modelling, AM enables better collaboration with the marketing department, while in the design review process, AM is used to perform fit and form testing of the given components. Finally, in production initiation, AM is used to fabricate fixtures to help in the debugging process, resulting in a cost saving on expensive machined fixtures.

Zebra Co., Ltd. has dominated the Japanese writing instruments manufacturing market. They implemented AM technologies, mostly for the purpose of new product development. Like other firms, the company had traditionally chosen outsourcing of its prototyping processes, which is not only a time-consuming and costly process, but the designers also lose the ability to rapidly modify and react to each design stage of the product development. Zebra used AM for the development of its innovative emulsion ink-based ballpoint pen, in which the smoothness of the writing experience is a key advantage of the pen, so that they could ensure the ease of grip in a minimum amount of time and at a low cost by using 3D printed prototypes that are very similar to the final products.

2.3.5 Other Industries

AM has been widely introduced in various industries, research and education centers, the arts and architecture, entertainment, machinery and equipment, and so on. The principal application sectors were discussed in previous sections, however, those cannot cover all the 3D printed markets. Thousands of case studies and examples can be made relating to all the available and attractive markets. However, we want to raise your attention by presenting some of those applications.

2.3.5.1 Architectural Models

Among the different potential applications, the most common use of AM for architecture includes the competition model and promotional models. It is possible to effectively present the idea to a jury or other professionals using a 3D printed competition model. AM is advantageous in this case because it can create a model regardless of its shape, even for one single unit. Promotional 3D printed models are a key component of virtual tours used to promote a real estate project. They allow investors or future clients to project themselves onto the area in question.

2.3.5.2 Food Industry

In the food industry and associated machinery, it can be a useful technology in the tooling and prototyping process, and even in the making of foods with a range of materials. For instance, AM technology is used to create the complex designs which are essential for machinery or complex and customized instruments used in its operation. Axon LLC manufactures the machinery used to shrink sleeve labels for containers. Most often the machines need custom tooling in order to shape the label to fit over the container. Many fixtures, called "pucks" are also required to hold the containers stable while the label is applied. Traditionally, the company produced prototype tooling, production tooling and pucks by machining them from plastic, which took approximately three weeks. The results of implementing AM in this company demonstrate a reduction of the production time from several weeks to only one day, and a production cost saving of $2400 for four pieces of tooling by a machining process, reducing down to $720 for the same parts made by 3D printing. The cost of printing a single puck is $12 compared to $125 for a machined one, a saving of about 90% on tooling costs. In addition, the manufacturer benefits from the low-cost design modification offered by AM, resulting in a better design process and final object.

Nestle Nespresso decided to buy a 3DP in 2003 in order to modernize the product development process of its coffee machines. The designers had traditionally outsourced the model making to external firms, which was a time-consuming process. It usually took several weeks to be prepared, thus any modification in design or any innovative changes were difficult to make. However, by using AM, they can now carry out any desired changes in just a few hours, something that will empower a more innovative approach—an essential factor for the development of innovative firms and products, something now possible thanks to the advent of AM. Nespresso engineers try out new ideas more often than ever before. The R&D manager said "Our creativity has grown by leaps and bounds, and we have gotten many ideas and results from the models that we created with the printer". In addition, the company, as other AM adopters have done, utilized AM technology for function testing of its models, producing handles, brewing units, capsule holders as well as other parts of its Nespresso machines as models to be functionally tested.

Companies also made different kind of foods based on the ordered design and available ingredients using 3DP. The Fab@Home team at Cornell University developed gel-like substances (food inks) called hydrocolloids that can be extruded and built up into different shapes. By mixing in flavoring agents, they can produce a range of tastes and textures (Pearce et al. 2010). Emerging 3D printing applications include food printers that can print in cake icing, cheese, or chocolate. The food products that are formed, such as company logos, names, and other unique objects will be widely printed. Any ingredients that exists in a liquid or powder form (or can be extruded through a nozzle or syringe) can be 3D printed. This includes sugar, cheese, sauces, and many others. An entire meal comprising of pizza, an eggplant dish, corn pasta, and pannacotta was recently 3D printed.

2.3.5.3 Research and Education

There are some studies (such as Eisenberg 2013; Groenendyk and Gallant 2013; Pryor 2014) that discuss the educational benefits of AM. Thanks to AM technologies, fabrication tools are becoming available to university undergraduates, school students, and even to children. Groenendyk and Gallant (2013) investigated the implication of 3DP and scanning at Dalhousie Libraries. Similar to the computers or Internet, AM technologies have the potential to change library services significantly. Not only does it allow physical items to be stored in a digital format, but it can also physically recreate these items on-demand for library patrons. 3D printing and scanning technologies can also provide advanced techniques of data visualization, enabling researchers to present data to audiences. Libraries can play an important role in this adoption and transition, much as they have in the past through the introduction of computers and the internet.

The Department of Industrial Design at Coventry University successfully employed AM aimed at the education sector. This department has reportedly gained popularity in educational leadership and development. The department utilizes AM technologies for a wide range of design applications, including fine art, jewelry, product design and automotive design. In 2006, the department installed the "Eden250", a 3-D printing system from Stratasys. The department's design technician stated that the principal benefits that AM brings to the educational process include significantly reducing production times and the promotion of creative thinking. It also affects the design process, because conventional manufacturing had limitations, while AM enables detailed designs that are more complex. AM allows for the production of parts that are already not feasible using conventional methods. In addition, they can now print one part instead of many parts that need to be assembled.

2.3.5.4 Pharmaceutical Industry

The pharmaceutical industry is also an AM application sector. For example, they used AM technologies to fabricate a custom-made daily pill for individual patients, which eliminates the need to keep track of multiple medications. AM is used for the controlled production of complex, multicomponent oral drug delivery tablets. Hopkinson et al. (2006) explained the process as follows: "A drug concentration profile can be generated in a computer model which may be produced by AM. By this means complex drug delivery regimes can be created featuring multiple drugs or the periodical, dosed release of a single drug." Identification markings on the pill eliminate the confusion and uncertainty of conventional medicines (Lipson 2012).

In general, three scenarios can be expected for the manufacture and distribution of on-demand 3D printed drugs (Lind et al. 2017), as follows:

- Factory, pharmacy (set dosage), patient
- Factory, patient (set dosage)
- Factory (set dosage), patient

The first scenario has a great impact on the distribution system of pharmaceuticals. In which, pharmacies are being compounded. They start with electronic prescriptions, which are a digital drawing and database of drug compounds, then the prescribed medicines can be 3D printed at a local pharmacy. In the second scenario, patients can use drug 3D printers, software, and/or apps to make their medicine at home, or the 3D drug printers can be remotely operated by healthcare professionals. In the third scenario, pharmaceutical firms will receive the patient's information and needs so that they can produce on-demand and customized drugs. However, the societal aspects of AM implementation in the pharmaceutical industry need to be taken into account. These aspects may involve regulations (for 3D printed drugs, for 3D printers) and the requalification of pharmacies.

Many other manufacturing sectors have introduced 3D printers to print things like jewelry and to make toys for kids, ultra light and perfectly fitted shoes, custom bike frames and road cycling helmets, and custom designed electric guitars, and so on. Table 2.2 details the significant AM application sectors, examples of current applications and the principal and commonly used materials. The output of this table provides a set of functional specifications that guides the implementation process.

Table 2.2 Overview of AM application sectors

Sectors	Current applications	Materials commonly used
Aerospace	Concept modeling and prototyping, structural and non-structural production parts, low volume replacement parts, (e.g., thrust reverser doors, landing gears, gimbal eye, fuel injection nozzles)	Titanium alloys, cobalt chromium alloys, stainless steels, nickel-based alloys, polyetherimide resins
Motor vehicles	Small quantities of structural and functional components (e.g., engine exhausts, drive shafts, gearbox components, and braking system for luxury) low volume vehicles (e.g., functional components for racing vehicles) low volume, (e.g., custom run speedometer housings, shrouds and fairings for motor cycles)	Titanium alloys, cobalt chromium alloys, stainless steels, ABS
Healthcare	Fabrication of custom-made prostheses and implants, medical devices, biological chips, tissue scaffolds, living constructs, drug-screening models, and surgical planning and training apparatus	Titanium alloys, cobalt chromium alloys, stainless steels, ABS, polyamides, nylon, photopolymers
Consumer products	I.e., Toys, figurines, furniture, office accessories, musical instruments, art, jewelry, museum displays, and fashion products	ABS, PC, SS, nylon, glass-filled polyamide, epoxy resins, wax and photopolymers cobalt chromium alloys
Industrial applications	Creation of end products that apply mechanical force to perform work	Titanium alloys, cobalt chromium alloys, stainless steels, nickel-based alloys, ABS
Architecture	Modeling of structures and designs	ABS, thermoplastic polymers
Government/military	For metal parts, heat exchangers, and use in remotely piloted vehicles.	Titanium alloys, cobalt chromium alloys, stainless steels, nickel-based alloys, polyetherimide resins

Adapted from uz Zaman et al. (2017)

Seven Key Facts

- AM products and services worldwide grew at a compound annual growth rate (CAGR) of 35.2% to $4.1 billion in 2014.
- AM can generate an economic impact of $230 billion to $550 billion per year by 2025, based on reducing costs and the value of customization.
- Approximately 67% of manufacturers are already using AM technologies.
- Industries such as consumer goods (hobbies and models, arts and fashions), automotive, aerospace, healthcare, and research institutions are the main AM application sectors.
- Aerospace industries employ AM technologies for concept models and prototypes, structural and non-structural production parts, and low volume replacement parts.
- Automotive industries employ AM technologies for small quantities of structural and functional components, low-volume vehicles (functional components for racing vehicles) and low production volume (custom run speedometer housings, shrouds and fairings for motor cycles).
- Healthcare industries employ AM technologies for custom-made prostheses and implants, medical devices, living constructs, drugs and surgical planning and training apparatus.

References

Achillas, C., Aidonis, D., Iakovou, E., Thymianidis, M., & Tzetzis, D. (2015). A methodological framework for the inclusion of modern additive manufacturing into the production portfolio of a focused factory. *Journal of Manufacturing Systems, 37,* 328–339.

Baur, C., & Wee, D. (2015). Manufacturing's next act. *McKinsey Quarterly*, Jun 2015.

Bogue, R. (2013). 3D printing: the dawn of a new era in manufacturing? *Assembly Automation, 33* (4), 307–311.

Boulton, C. (2013). Printing out barbies and Ford cylinders. *The Wass Street Journal*, Dow Jones & Company, Inc. http://www.wsj.com/articles/SB10001424127887323372504578469560282 127852.

Campbell, I., Bourell, D., & Gibson, I. (2012). Additive manufacturing: Rapid prototyping comes of age. *Rapid Prototyping Journal, 18*(4), 255–258.

Campbell, T., Williams, C., Ivanova, O., & Garrett, B. (2011). *Could 3D printing change the world? Technologies, potential, and implications of additive manufacturing.* Washington, DC: Atlantic Council.

Cozmei, C., & Caloian, F. (2012). Additive manufacturing flickering at the beginning of existence. *Procedia Economics and Finance, 3,* 457–462.

Eisenberg, M. (2013). 3D printing for children: What to build next? *International Journal of Child-Computer Interaction, 1*(1), 7–13.

Ford, S. L. (2014). Additive manufacturing technology: Potential implications for US manufacturing competitiveness. *Journal of International Commerce and Economics*, September 2014. Available at SSRN: https://ssrn.com/abstract=2501065

Freedman, D. (2011). Layer by layer. *MIT Technology Review*. Retrieved from: https://www.technologyreview.com/s/426391/layer-by-layer/, December 19, 2015.

Giannatsis, J., & Dedoussis, V. (2009). Additive fabrication technologies applied to medicine and health care: A review. *International Journal of Advanced Manufacturing Technology, 40*(1), 116–127.

Gibson, I., Rosen, D. W., & Stucker, B. (2010). *Additive manufacturing technologies*. New York: Springer.

Goffin, J., Van Brussel, K., Martens, K., Vander Sloten, J., Van Audekercke, R., & Smet, M. H. (2001). Three-dimensional computed tomography-based, personalized drill guide for posterior cervical stabilization at C1–C2. *Spine, 26*(12), 1343–1347.

Groenendyk, M., & Gallant, R. (2013). 3D printing and scanning at the Dalhousie University libraries: A pilot project. *Library Hi Tech, 31*(1), 34–41.

Guo, N., & Leu, M. C. (2013). Additive manufacturing: Technology, applications and research needs. *Frontiers of Mechanical Engineering, 8*(3), 215–243.

Hasan, S., Rennie, A. E. W. (2008). The application of rapid manufacturing technologies in the spare parts industry. In *Nineteenth Annual International Solid Freeform Fabrication (SFF) Symposium*, August 4–8, 2008, Austin, TX, USA.

Holmström, J., Partanen, J., Tuomi, J., & Walter, M. (2010). Rapid manufacturing in the spare parts supply chain: Alternative approaches to capacity deployment. *Journal of Manufacturing Technology Management, 21*(6), 687–697.

Hopkinson, N., Hague, R., & Dickens, P. (Eds.). (2006). *Rapid manufacturing: An industrial revolution for the digital age*. Wiley.

IndustryWeek, (June 2013). *The next wave of manufacturing*. Retrieved from http://www.industryweek.com/technology/next-wave-manufacturing-human-organs.

Khorram Niaki, M., & Nonino, F. (2017). Impact of additive manufacturing on business competitiveness: A multiple case study. *Journal of Manufacturing Technology Management, 28*(1), 56–74.

Kochan, A. (2003). Rapid prototyping helps Renault F1 Team UK improve championship prospects. *Assembly Automation, 23*(4), 336–339.

Lind, J., Kälvemark Sporrong, S., Kaae, S., Rantanen, J., & Genina, N. (2017). Social aspects in additive manufacturing of pharmaceutical products. *Expert Opinion on Drug Delivery, 14*(8), 927–936.

Lipson, H. (2012). Frontiers in additive manufacturing. *Bridge, 42*(1), 5–12.

Lipson, H., & Kurman, M. (2013). *Fabricated: The new world of 3D printing*. Wiley.

Liu, P., Huang, S. H., Mokasdar, A., Zhou, H., & Hou, L. (2014). The impact of additive manufacturing in the aircraft spare parts supply chain: Supply chain operation reference (SCOR) model based analysis. *Production Planning & Control, 25*(13–14), 1169–1181.

Liu, L., Xiong, Z., Yan, Y., Zhang, R., Wang, X., & Jin, L. (2009). Multinozzle low-temperature deposition system for construction of gradient tissue engineering scaffolds. *Journal of Biomedical Materials Research. Part B, Applied Biomaterials, 88*(1), 254–263.

Manyika, J., Chui, M., Bughin, J., Dobbs, R., Bisson, P., & Marrs, A. (2013). *Disruptive technologies: Advances that will transform life, business, and the global economy* (Vol. 12). San Francisco, CA: McKinsey Global Institute.

Pearce, J. M., Blair, C. M., Laciak, K. J., Andrews, R., Nosrat, A., & Zelenika-Zovko, I. (2010). 3-D printing of open source appropriate technologies for self-directed sustainable development. *Journal of Sustainable Development, 3*(4), 17.

Petrovic, V., Vicente Haro Gonzalez, J., Jorda Ferrando, O., Delgado Gordillo, J., Ramon Blasco Puchades, J., Portoles Grinan, L. (2011). Additive layered manufacturing: Sectors of industrial application shown through case studies. *International Journal of Production Research, 49*(4), 1061–1079.

Prototype Today (February 2013). *3D printer creates ear using injectable gels.* Retrieved from http://www.prototypetoday.com/weill-cornell-medical-college/3d-printer-creates-ear-using-injectable-gels-made-of-living-cells.

Pryor, S. (2014). Implementing a 3D printing service in an academic library. *Journal of Library Administration, 54*(1), 1–10.

Quadrani, P., Pasini, A., Mattioli-Belmonte, M., Zannoni, C., Tampieri, A., Landi, E., et al. (2005). High-resolution 3D scaffold model for engineered tissue fabrication using a rapid prototyping technique. *Medical & Biological Engineering & Computing, 43*(2), 196–199.

Reeves, P. (2008). How rapid manufacturing could transform supply chains. *Supply Chain Quarterly, 2*(04), 32–336.

Rengier, F., Mehndiratta, A., von Tengg-Kobligk, H., Zechmann, C. M., Unterhinninghofen, R., Kauczor, H. U., et al. (2010). 3D printing based on imaging data: Review of medical applications. *International Journal of Computer Assisted Radiology and Surgery, 5*(4), 335–341.

Rochus, P., Plesseria, J. Y., Van Elsen, M., Kruth, J. P., Carrus, R., & Dormal, T. (2007). New applications of rapid prototyping and rapid manufacturing (RP/RM) technologies for space instrumentation. *Acta Astronautica, 61*(1), 352–359.

Ruffo, M., & Hague, R. (2007). Cost estimation for rapid manufacturing simultaneous production of mixed components using laser sintering. *Proceedings of the Institution of Mechanical Engineers, Part B: Journal of Engineering Manufacture, 221*(11), 1585–1591.

Shah, S. K., & Tripsas, M. (2007). The accidental entrepreneur: The emergent and collective process of user entrepreneurship. *Strategic Entrepreneurship Journal, 1*(1–2), 123–140.

Singh, S., Ramakrishna, S., & Singh, R. (2017). Material issues in additive manufacturing: A review. *Journal of Manufacturing Processes, 25*, 185–200.

Stevens, B., Yang, Y., Mohandas, A., Stucker, B., & Nguyen, K. T. (2008). A review of materials, fabrication methods, and strategies used to enhance bone regeneration in engineered bone tissues. *Journal of Biomedical Materials Research. Part B, Applied Biomaterials, 85*(2), 573–582.

uz Zaman, U. K., Siadat, A., Rivette, M., Baqai, A. A., & Qiao, L. (2017). Integrated product-process design to suggest appropriate manufacturing technology: A review. *International Journal of Advanced Manufacturing Technology, 91*(1–4), 1409–1430.

Vandenbroucke, B., & Kruth, J. P. (2007). Selective laser melting of biocompatible metals for rapid manufacturing of medical parts. *Rapid Prototyping Journal, 13*(4), 196–203.

Walter, M., Holmström, J., Tuomi, H., Yrjölä, H. (2004, September). Rapid manufacturing and its impact on supply chain management. In *Proceedings of the logistics research network annual conference* (pp. 9–10).

Wohlers, T. (2012). *Wohlers report 2012.* Wohlers Associates. Inc., ISBN 975442988.

Wohlers, T. (2014). *Wohlers report 2014: 3D printing and additive manufacturing state of the industry.* Fort Collins, CO, USA: Wohlers Associates Inc.

Wohlers, T. T., & Caffrey, T. (2015). Wohlers report 2015: 3D printing and additive manufacturing state of the industry annual worldwide progress report. *Wohlers Associates.*

World Economic Forum (2015). *Deep shift technology tipping points and societal impact.* Survey Report of Global Agenda Council on the Future of Software & Society.

Chapter 3
The Value for Sustainability

Forced by competition, by international and national politics and by the "green paradigm", in recent decades managers have been increasingly considering the sustainability aspects of value creation. Sustainability, or perhaps better, sustainable development "involves the simultaneous pursuit of economic prosperity, environmental quality, and social equity" (Elkington 2002) while sustainable manufacturing (see Sect. 4.2) refers to the creation of manufactured products that use processes that are non-polluting, conserve energy and natural resources, and are economically sound and safe for employees, communities, and consumers alike. In an integrational perspective, the key characteristic is the integration of economic, environmental, and social aspects, and the relationships between them (Lozano 2008) (Fig. 3.1).

Several characteristics can be representative of the future uncertainties for sustainable development, such as the unpredictability of fossil fuel supplies and associated price fluctuations, use of non-renewable raw materials, shortage of raw materials, global warming due to environmental pollution, and environmental protection from chemicals, limited space availability, and higher landfill taxes. The emergence and development of AM technologies can address all of these aspects. According to the study by Gebler et al. (2014), AM technologies will have huge potential sustainability implications on a product's lifecycle, considering the cost, energy and CO_2 emissions. This amounts to cost reductions of 170–593 billion US \$, avoided Total Primary Energy Supply (TPES) of 2.54–9.30 exajoules (EJ) and avoided CO_2 emissions of 130.5–525.5 Metric tons (Mt) by 2025 in the markets identified for 3D printing. However, the influence of AM on the global economy, energy consumption, and environments is expected to gradually increase over decades, as has been the case with the internet and personal computers.

This chapter discusses the value of AM for sustainability. It begins with the explanation of its values for economy and goes on to describe the impacts for the environment, including respectively the impact on energy consumption, and the impact on resource usage and pollution. Furthermore, the chapter discusses the social implications of AM, its advantages and the challenges that lie ahead.

© Springer International Publishing AG 2018   67
M. Khorram Niaki and F. Nonino, *The Management of Additive Manufacturing*,
Springer Series in Advanced Manufacturing,
https://doi.org/10.1007/978-3-319-56309-1_3

Fig. 3.1 Integrated view of sustainable development

3.1   Economic Sustainability

In today uncertain economic situation, AM is a helpful technology for companies that need to survive and be competitive. In a similar manner to the time when Kodak digital camera caused the decline of film-based camera, likewise in the case of conventional manufacturing a big shift could be caused by additive technologies. In order to be competitive against low-cost mass production from overseas, companies have to focus on innovative and ever more complex goods, which are usually customized based on the needs of small consumer groups or are unique to an individual. Various researchers have demonstrated that technological innovation influences both firms and market structures.

AM industries and application sectors are considered to be a 230–550 billion US $ market by 2025 (McKinsey Global Institute 2013), in which the main economic impacts are linked to markets with high value, low volume and customized products.

3.1.1   *Operational and Manufacturing Costs Savings*

Studies demonstrate a huge amount of expected saving with the adoption of AM technologies. For instance, Gebler et al. (2014) estimated that the savings will be in the order of 113–370 billion US$ by 2025, which comes from savings in material inputs and operational costs, and through the shortening of the supply chain. The analysis includes all the sustainability implications of AM technologies throughout the product lifecycle, consisting of raw materials, manufacturing, distribution in the production phase as well as the use phase. Sustainability aspects dealing with cost are discussed in this section, and other aspects including energy consumption and CO_2 emissions will be discussed in subsequent paragraphs. Four scenarios are established, since both market potential and process intensities are associated with uncertainties. In addition, five leading AM application sectors are involved in this analysis, namely consumer products, aerospace, automotive, medical industry and tooling.

Figure 3.2 shows quantified sustainability implications of lifecycle through AM, considering the operational cost components. From the figure, it can be seen that saving potentials are mainly during the production and use phase of a product. Varying with the scenarios, cost savings in the production phase is expected to be 113–370 billion US$, and 56–219 billion US$ in the use phase, all by 2025. In total, by 2025, it is expected to reach 170–593 billion US$ over the entire lifecycle.

This large amount of cost savings is thanks to the advent of additive technologies, which suggest the practitioners to use the technology. The majority of cost savings in production phase are due to reduced handling, a shorter supply chain and fewer demands for materials. AM technology also results in cost savings in the

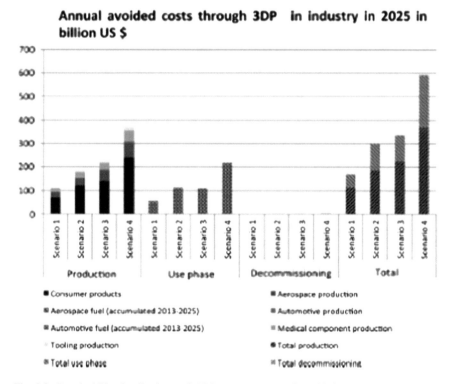

Fig. 3.2 Sustainability implications of AM on costs: annual avoided costs through AM in industry. Adapted from Gebler et al. (2014)

production phase through on-demand production. This flexible production approach has benefits for manufacturers having a minimal need for warehousing raw materials, work in progress and final end-use products. It also needs a minimum of labor for pre- and post-processing, since the function is highly automated. In the use phase of product, about one-third of the total savings is obtained from lightweight parts, resulting in lower fuel consumption.

The first application sector to benefit from cost savings is the consumer products sector, and the second is aerospace industries. In aerospace industries, savings in fuel consumption in use phase, is the most advantageous impact of AM technologies, since fuel demand is highly influenced by a component's weight. The relative largest influence of AM in cost saving occurs in aerospace fuel demand, followed by the medical industry, however, automotive and consumer products were the larger markets for potential AM impacts, the relative influence remaining low. This has arisen for several reasons; firstly, the production volumes are usually higher in these sectors; secondly, the process speed in AM is lower than that in conventional manufacturing; and thirdly because of the limitation in material ranges, it will be difficult to replace all products.

In the past decades, Flexible Manufacturing Systems (FMS) has offered numerous advantages for operations. FMS is a manufacturing system that can produce a variety of components using same resources at the minimum changing cost. AM technologies have extended the features of the conventional FMS. It can be considered actually as a novel and advanced FMS system. Likewise the FMS and certainly more than its capabilities, AM affects the costs particularly the cost of flexibility, customization, new product development, capital, and marginal production costs. In contrast, economy of scales is the point that these two manufacturing systems are distinguished. Economies of scale are the cost compensations that arise with the increased output of a production. Economies of scale arise from the inverse relationship between the quantity of production and fixed-costs per unit. In other words, the greater the quantity of production, the lower the per-unit fixed cost will be, since these costs are spread out over a larger number of goods. In larger production volume, AM becomes less competitive, while the costs per unit of conventional FMS decrease. Accordingly, AM become more competitive, for lower production volume where the costs per unit of conventional FMS increase.

Figure 3.3 illustrates the comparison between AM and FMS systems relating to economies of scale. As can be seen from the figure, the larger the production volume, the less beneficial AM becomes in terms of cost (the cost curve for conventional FMS is adjusted downward as illustrated in Fig. 3.3a). Consequently, at a definite level of economies of scale, the advantage of the AM entrant would be almost nothing, which is the high economy of scale point. In Fig. 3.3b, other factors aside from production volume were considered, since production volume does not involve the differences between the outputs, which can be the result of the quality of the products and order fulfilment times. As seen in the figure and based on the study of Weller et al. (2015), the incentives to adopt AM technologies are low when the requirements for time and quality are too high.

(a)

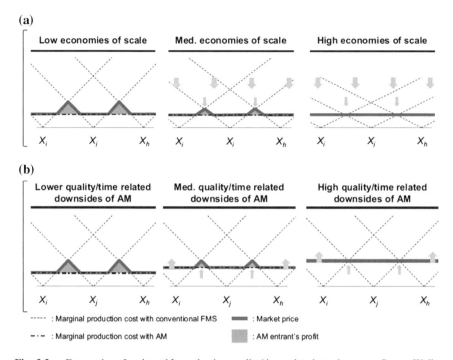

(b)

Fig. 3.3 **a** Economies of scale and **b** production quality/time-related requirements. *Source* Weller et al. (2015)

Using additively manufactured parts can also lead to cost savings in the use phase of the products' lifecycle, through lightweight components that will reduce raw material and energy consumption and could deliver savings of 56–219 billion USD by 2025 (Gebler et al. 2014). Also cost savings in the use phase, can be obtained from the biological production principle, which involves layer-by-layer approaches that are inherently less wasteful than traditional subtractive methods of production. Additionally, the capability of producing integrated parts without the need for assemblies has profound effects on the production line. Further to its quality impact, integrated products will lead to the minimizing of waste and correspondingly lower labor costs. It also facilitate a flatter and simpler supply chain thanks to the requirement for fewer components.

Additionally, both the production process and design process based on AM is more efficient in terms of energy and resources used. A summary of the potential sustainability effects of AM and both benefits and challenges is presented in Table 4.2 of Chap. 4.

3.1.2 Logistics' Costs Saving

AM overlay enables both small and large enterprises to compete within the global economy, and it is impactful at all stages of business, from primary product development, throughout manufacturing, to logistics, and after sale services.

As regards logistics cost, the adoption of AM technology will promote the shifting of the production location nearer to—or even at—the point of final consumers, resulting in geographically delocalized production. In addition, because the manufacturers can be dispersed or may be in multiple locations, the single source supply risk will therefore be diminished. More importantly, the transportation costs and corresponding environmental aspects can be easily minimized with an optimum production planning and location allocation strategy. Therefore, looking at the sustainability impacts of AM on logistics and the supply chain, it mostly seems beneficial for reconfiguring the value chain to include a shorter supply chain, providing localized production to reduce inventory, and enabling the introducing of novel distribution models to realize efficient transportation planning. The detailed discussions on supply chain and business model for consumer products—which are linked to sustainability issues—are provided in Sect. 4.2.

3.1.3 Implications for Corporate Taxation

Further to evaluating the potential impacts of AM in terms of improving efficiency, some other economic aspects should be taken into account. Cozmei and Caloian (2012) investigated the implications and consequences of AM for business, accounting, and tax, rather than relying on the application of predefined rules. Firstly, since AM meets the requirements for sustainable development, for instance in reducing the use of non-renewable resources, reducing the waste production and carbon emissions, the environmental protection expenditure and associated taxes will collapse. Secondly, AM technologies enable the reducing of operational costs, and it affects import and export flows in the long term by not requiring centralized production in one location, and it also needs less labor, which all lead a lower tax burden. Furthermore, as a result of the smaller workforce, charges for social insurance systems will drop considerably, even though a challenging question asks whether these reduced taxes will be able to uphold the healthcare and pension systems.

Currently for fiscal reasons, the adoption of such a technology would involve only benefits since research and design activities are stimulated (expenses are fully deductible in the calculation of taxable income), and the profit invested in the acquisition of AM equipment could be exempt from taxation, etc. However, the virtual content has enjoyed less tax certainty. Governments will have to consider this issue and address tax guidelines to improve tax transparency and tax treatment in the virtual business sphere, because the adoption of AM creates a new business model that do not fit into a traditional model of taxation. In the case of virtual goods, purchasers receive access codes on a printed receipt, allowing them to access the content on a third-party server over the internet. The content downloaded should be treated as taxable. Digital files could be compared to digital goods: software, music or other electronic files that buyers download exclusively from the internet. To sum up the tax issues, there should be a proper identification and definition of a "taxable event" in a virtual world or virtual economy, so that a new tax policy can be shaped (definition and realization of income, recognition events, and characterization). For practical considerations, a tax system for the virtual economy could be implemented, because this environment has the advantage that it can be controlled.

3.2 Environmental Sustainability

The environmental aspects of AM will form one of the major pillars of its development over the next ten years (Bourell et al. 2009) that should certainly be taken into account. Environmental assessments of AM guide manufacturers to either identify the impacts through comparing the technologies and making a decision in order to select a technology, or to find the potential optimization points to enhance the value of implementation. There are several studies assessing the impacts of AM technologies on the environment, and these studies can be divided into three subgroups, compromising comparisons of various AM technologies and comparisons between AM and conventional manufacturing methods in terms of energy consumption, optimization of that energy consumption and an assessment of the other environmental aspects of AM.

3.2.1 Impacts on Energy Consumption

Several studies compared the energy consumption of different AM technologies and between AM and conventional manufacturing methods. The specific energy consumption (SEC) of AM processes is approximately 100-fold higher than that of conventional bulk forming processes (Yoon et al. 2014). Nevertheless, as with small production volumes, fused deposition modeling (FDM) has a greater advantage compared to conventional manufacturing processes as seen in Fig. 3.4.

The study of Yoon et al. (2014) compared the energy consumption of different manufacturing methods. It employed various methods including inject molding bulk-forming processes, milling from a subtractive process and FDM from an additive process. The size of the sample part was $30 \times 30 \times 5$ mm, and the material used was ABS P400. As for the inject-molding process, the total energy consumption for one standard sample was 832.4 Wh, or 222.0 kWh kg^{-1}, comprising 189.3 Wh for mold fabrication and 606.1 Wh for the warm-up stage.

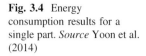

Fig. 3.4 Energy consumption results for a single part. *Source* Yoon et al. (2014)

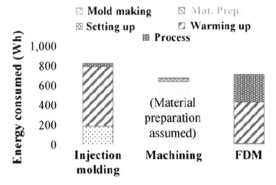

However, when the number of parts to be fabricated increased, the SEC decreased, because the processing energy accounted for only 9.9 kWh kg^{-1}. In the milling process, the total energy consumption was 40.8 Wh per part, or 10.9 kWh kg^{-1}. Finally, the part that was produced by the FDM process, consumed 717.1 Wh per part, or 191.3 kWh kg^{-1}. However, in this case, the specific energy consumption will also decrease when the number of parts to be produced increases, because the constant energy consumption of the warm-up process was about 435.5 Wh. The summary of the results and the share of each manufacturing stage such as mold making, setting up, and processes in inject molding, material preparation and warming up in the FDM process are depicted in Fig. 3.4.

Consequently, as the study reveals, the conventional bulk-forming process has the lowest range of SEC and the additive process has the highest. Therefore, injection molding and machining were highly suitable techniques for mass production. Even so, when the number of parts was one, bulk forming consumed the largest amount of energy, in comparison with AM and machining. Therefore, FDM may be more efficient for single or low volume production (see Fig. 3.5).

Regarding production-related energy consumption, studies demonstrate the efficiency of conventional manufacturing (injection molding) rather than additive process (SLS) for large production volumes (e.g. Chen et al. 2015). Figure 3.6 illustrates a trade-off analysis between these two manufacturing methods. The starting point on the y-axis is the energy input of the pre-chain process, for instance the molds required for the injection molding process. The slope is the energy intensity of the processes and the proportion of the material input to the weight output of the final product. SLS systems are better for small production volumes. In fact, there are no economies of scale at the SLS energy demand. With this system, the energy demand is constant even for larger production volumes, nevertheless as with injection molding it is significantly decreasing.

In general, when looking at energy consumption, AM is usually not as efficient as conventional manufacturing (Khorram Niaki and Nonino 2017b). Table 3.1 shows the Specific Energy Consumption (SEC), which is defined as the energy consumption in the production of a material unit. These measures are representative only at the individual processing level. The FDM process usually uses heaters to

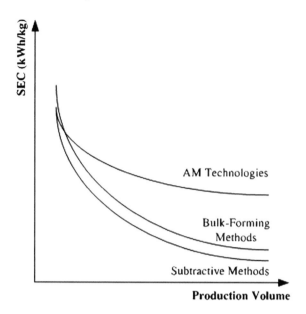

Fig. 3.5 Specific Energy Consumption (SEC) comparison of production volumes

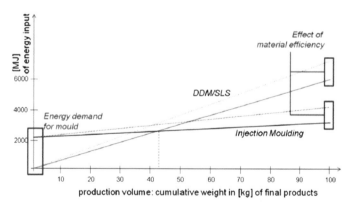

Fig. 3.6 A comparative energy trade-off analysis between injection molding (IM) and selective laser sintering (SLS). *Source* Chen et al. (2015)

maintain the temperature of the chamber to ensure a stable process, and this causes the higher SEC of laser sintering processes such as SLS and SLA, however, FDM has a lower energy intensity compared to the others.

In contrast, Baumers et al. (2013) showed the efficient energy consumption of AM technologies when considered in terms of the optimization of feasible design (see Sect. 7.3.2 for detail) in comparison with conventional manufacturing methods. Moreover, the adoption of AM may result in a restructuring of the value chain

Table 3.1 *SEC* comparison between bulk-forming, subtractive, and additive processes (own table based on the study of Yoon et al. 2014)

Bulk-forming processes	Process	Metal casting			Injection molding		
	Detail	Different material			Electric method	Hybrid method	Hydraulic method
	SEC (kWh kg^{-1})	0.62–6.89			0.11–1.27	0.47–2.04	0.19–5.82
Subtractive processes	Process	Drilling		Turning		Milling	
	Material	Grey cast iron		Brass	Steel	Aluminum	Steel
	SEC (J mm^{-3})	9–65		2.7–9.8	5.3–36.2	2.3–4.9	6.8–151.8
Additive processes	Process	EBM	SLM	DMLS	3DP	SLA	FDM	SLS
	SEC (kWh kg^{-1})	17–49.17	27–31	24.2–94.17	14.7–17.4	20.70–41.38	23.08–163.69	14.5–66.02

through the redesigning of the products and components into simpler products that require fewer components, materials, stages and interactions. Therefore, with respect to the value chain obtained by the redesigned products, the overall environmental impacts over the whole supply chain would be affected, since the scale of material flows have been reduced. Therefore, AM can enable the reduction of energy consumption through operations outside of the manufacturing processes such as reducing or eliminating transportation through distributed manufacturing, and inside the operations through reducing material waste, providing re-manufacturability and an optimized design.

In addition, the potential for process redesign offered by AM encompasses many application sectors, for instance the construction sector, which is a major material, energy and water consumer. Therefore, it provides major opportunities for resource efficiency improvements. Developing in situ construction (production on site and in the original place of use), such as the MX3D bridge and 3D printed apartment buildings depicts what is becoming feasible using AM technologies in this industry. The MX3D is going to 3D print a fully functional steel bridge over water in the center of Amsterdam. Although the use of AM in this industry is in its infancy, the preliminary cases demonstrate the reduction of the environmental impacts of logistics and the use of basic materials meaning that fewer materials can be used and there are fewer waste overheads. However, the construction industry is highly resistant and slow to change. This resistance to the adoption of AM technologies arising from the longer life cycle of infrastructure and concerns about safety and liability. Therefore, experiments and evidence of the infrastructure over a longer timeframe are needed, since the product lifecycles are shorter in other industries compared to those in construction.

The cost implications of the product lifecycle of 3D printed markets based upon the results of Gebler et al. (2014) were discussed in Sect. 3.1. As mentioned earlier, the study also considered other sustainability implications of AM technologies including energy and CO_2 emissions in a product's lifecycle. In terms of energy, Fig. 3.7 depicts the quantified sustainability implications through AM.

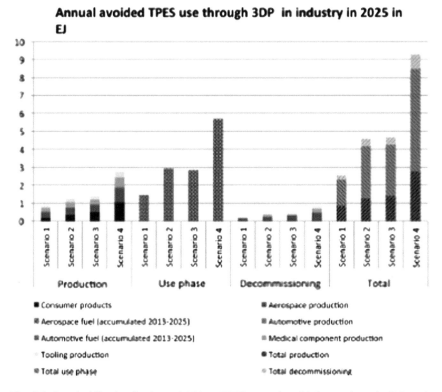

Fig. 3.7 Sustainability implications of AM on TPES: annual avoided costs through AM use in industry. *Source* Gebler et al. (2014)

Sustainability implications of energy were considered in terms of the total primary energy supply (TPES) and whether the primary energy is non-renewable or renewable. The TPES is a measure that indicates the sum of production and imports subtracting exports and storage changes. Figure 3.7 shows TPES savings through AM over the entire lifecycle of a product. In the production phase, TPES savings are 0.85–2.77 EJ, and in use phase, savings are 1.46–5.72 EJ depending on the scenario in 2025. EJ is a unit of energy equal to 1.055×1018 J (1.055 exajoules or EJ) in SI units, or 1015 (a short-scale quadrillion) British Thermal Units (BTU)—the unit used in discussing world and national energy budgets.

In general, total reduction in TPES through AM is expected to be 2.54–9.30 EJ over the entire product lifecycle. Approximately, one-third of the TPES savings occur in the production phase, 55–60% of the total savings can be harvested during the use phase and 8% are obtainable during the decommissioning phase. The most beneficial application sectors for reducing TPES are consumer products and the aerospace industry. Usage-phase TPES savings occur exclusively to aerospace energy demands due to the abovementioned reason for costs. The relative impact on the TPES of a market occurs respectively to aerospace fuel demands (9–35%), aerospace production (8–19%), medical components (5–19%) and tooling (3–10%). AM does not contribute to the large relative influence on consumer products and automotive industries, since these markets involve mass production, in which AM technologies are not yet cost effective and energy effective.

3.2.2   Impacts on Resources and Pollutions

Several studies assessed not only energy consumption, but also all existing environmental aspects, providing insights into the sustainability of AM in terms of energy/resources/time consumption, and greenhouse gas emission, hazard, and accuracy (e.g., Drizo and Pegna 2006; Le Bourhis et al. 2013; Sharif Ullah et al. 2013). They developed a sustainability index for AM processes, which includes sustainability factors such as the volumetric quantity of model-building material, as well as carbon footprint and resource depletion of the primary production of model-building materials, energy consumption, and carbon dioxide emissions of the model-building process. Some of the selected studies relating to the pollution are summarized below.

Various methodologies have been developed in order to assess environmental aspects. Lifecycle Assessment/Analysis (LCA), developed by the Society of Environmental Toxicology and Chemistry (SETAC), has gained the greatest popularity. It takes into account all the stages of a product's life from raw materials to manufacturing, distribution, usage, and recycling. In fact, it enables the accurate and comprehensive quantification of the environmental aspects of a global system. There are four linked components in the LCA method (Fig. 3.8):

- Goal and scope: identifying the purpose of the study and its boundaries
- Lifecycle inventory: identifying all the input flows, including energy and raw materials, etc. and environmental releases associated with each stage of a lifecycle

Fig. 3.8 Phases of lifecycle assessments

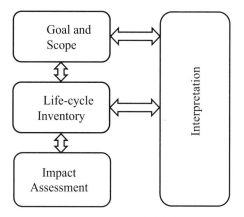

- Impact assessment: assessing the identified impacts using category indicators such as Eco-Indicator 99
- Interpretation: drawing conclusions, checking for completeness, contribution, sensitivity analysis, and finding opportunities for improvements.

Initially, the parameters of the environmental aspects of AM should be identified. To this end, Le Bourhis et al. (2013) studied the environmental impacts of an AM process. As an example of a specific AM process, the environmental aspects of the Direct Metal Laser Sintering (DMLS) process have been identified. These include three main flows in production cycle—fluids, raw materials, and energy consumptions. There are two different kinds of fluid consumption. Firstly, the inert gas used to carry powder from the bowl to the melting bed and to form a perfect powder cone. The second fluid is the hydraulic fluid, used to cool down the laser unit and the axes of the machine.

With materials, although the AM machine exclusively fuses the necessary powders, an amount of powder will not be fused. As for material consumption estimation, different parameters are considered, taking into account material characterization factors, nozzle efficiency, powder flow rate, manufacturing time, and the weighting factor. The latter allows the accounting of the weight of lost powder comparing to fused powder. In addition, as with the nozzle, it should be recognized that that the size of the nozzles and the desired powder flow rate affect the efficiency. In fact, a smaller nozzle has a lower efficiency. In general terms, AM technologies promote efficient material usage when compared to conventional manufacturing methods, however, unskilled operators and a poor design may still result in the creation of waste. For instance, Fig. 3.9 shows the possible errors in the FDM process, causing waste throughout the production process.

In electricity consumption, there are some constant energy consumptions during a process, for example, by some hydraulic components or electrical cabinets in DMLS machines, whereas with the others, the energy consumption depends on the process such as laser, cooling system, and motor axes. The study of the DMLS process also considers the design optimization process offered by AM. In fact, their methodology, on the one hand, enables the proposition of a minimization of the environmental impacts through either design modification, and/or process parameters optimization. On the other hand, the methodology allows the determining of the share of each flow consumption (electricity, fluids, material) in the global environmental impact. It was argued that it is not judicious to focus only on the electricity consumption, and other flows such as materials and fluids have to be considered.

Table 3.2 shows comparative measures of the environmental impact of the various AM systems available. In the table, the Eco-Indicator 95 method has been used. This is a European scale method for the weighting of environmental impacts that damage ecosystems. Eco-Indicator 95 contains 100 indicators relating to important materials and processes. ECR (kWh/kg) is the massive energy use during the process and the environmental impact was calculated based upon the multiplication of ECR and constant measures (= 0.57 millipoints (mPts)/kWh).

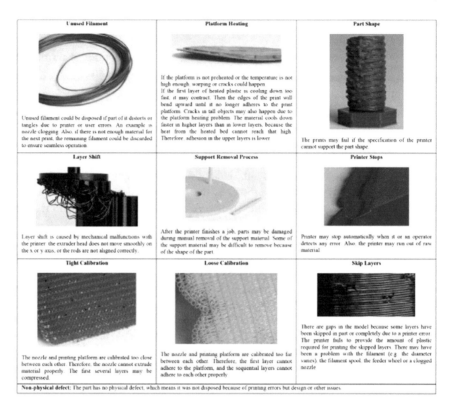

Fig. 3.9 Reasons for waste creation in fused deposition modeling (FDM). *Source* Song and Telenko (2017)

Therefore, the final measure will be an environmental impact per kilogram expressed in millipoints per kilogram (mPts/kg).

For instance, Luo et al. (1999) evaluated the SLA process. The study considered the epoxy resin as raw materials, and compared three different types of SLA machines (the SLA 250, SLA 3500, and SLA 5000). They took into account several parameters such as raw material usage, speed of the scanning process, line width, the power rate of the machine, and the process delay between layers. The results revealed that the SLA 250 had a smaller ECR than the SLA 3500, with the SLA 5000 having the smallest ECR, and consequently the lowest environmental impacts. Likewise, for the SLS process and polyamide material, the EOSINT P760 and Vanguard HiQ had a smaller ECR than the others, and the Sinterstation DTM 2000 had the largest ECR and consequently the largest environmental impacts.

AM has the potential to reduce CO_2 emissions over the entire lifecycle of a product. The CO_2 emissions are coherent with total primary energy supply as they relate directly to energy. These reductions are expected to be in the order of 34.3–151.1 Mt in the production phase, 84.1–328.5 Mt in the usage phase and 12.1–44.5 Mt in the decommissioning by 2025 (Gebler et al. 2014). In broad terms, the

Table 3.2 Environmental impacts of different AM machines using Eco-Indicator 95

AM systems	Machines	Materials	ECR (kWh/kg)	Environmental impact (mPts/kg)
Stereolithography	SLA-250	Epoxy resin SLA 5170	32.48	18.51
	SLA-3000	Epoxy resin SLA 5170	41.41	23.60
	SLA-5000	Epoxy resin SLA 5170	20.70	11.8
Selective laser sintering	Sinterstation DTM 2000	Polyamide	40.01	22.81
	Sinterstation DTM 2500	Polyamide	29.77	16.97
	Vanguard HiQ	Polyamide	14.54	8.29
	EOSINT M250	Metalic powder (Bronze + Ni)	5.41	3.09
	EOSINT P760	Polyamide PA2200 Balance 1.0	36.50	20.81
	EOSINT P760	Polyamide PA2200 Speed 1.0	39.80	22.69
	EOSINT P760	Polyamide PA3200GF	26.30	14.99
Fused deposition modeling	FDM 1650	ABS Plastic	346.43	197.47
	FDM 2000	ABS Plastic	115.48	65.82
	FDM 8000	ABS Plastic	23.10	13.16
	FDM Quantum	ABS Plastic	202.09	115.19
Selective laser melting	MTT SLM 250	Metallic powder SAE 316L	31.00	17.67
Electron beam melting	Arcam A1	Metalic powder Ti-6AI-4V	17.00	9.69

Source Le Bourhis et al. (2013)

total savings in CO_2 emissions through AM are expected to be 130.5–525.5 Mt in 2025 over the entire lifecycle of the product. The production phase has more than one quarter of the total saving potentials, however, the largest savings are achievable in the usage phase due to the high energy saving potential of the aerospace industry through increasing of the strength-to-weight ratio. The relative impact on CO_2 emissions (same as TPES) of a market occurs as follows: on aerospace fuel demands (9–35%), aerospace production (8–19%), medical components (5–19%) and tooling (3–10%). Consumer products and the automotive industry contribute little, for the reasons mentioned earlier.

As mentioned before, to promote the selection of the AM process as a manufacturing method, an accurate and comprehensive sustainability comparison with conventional manufacturing will help manufacturers make an efficient decision. Sharif Ullah et al. (2013) conducted an experiment with the fabrication of a given model (prototype) once with AM (using an SLA process) and once with a subtractive CNC machining process (wooden material-based). The physical models were made using the same 3D CAD data. In order to assess the environmental impacts and compare them with those of conventional manufacturing, a sustainability index was used, incorporating CO_2 footprint and resource depletion of primary production of model-building material, energy consumption and the CO_2 emissions of the model-building process. In the comparative study, it was revealed that an AM process uses less material, because of its capability of producing hollow objects. However, SLA consumes more energy and requires a longer building time. The study also argues that the CO_2 footprint of initial material production for the wooden material-based prototyping is better than SLA. However, considering resource depletion (i.e., materials and water usage), the SLA process is better than the wooden material-based RP technology.

In spite of the all the potential to reduce waste and increase resource efficiency through the use of AM, its printable materials are not necessarily greener than the common materials used in conventional manufacturing. According to Faludi et al. (2015) the one exception is the bio-polymer polylactic acid (PLA). In addition, the relative toxicity of the materials and the energy usage in producing the input material are the entire environmental aspects of the system that should be taken into account. In fact, considerable energy is consumed in the processing or refining of the metal ores in preparation for the AM process. During this processing stage, there is the potential to rethink how raw materials should be processed to minimize the resources required to render them suitable as inputs for additive processes. The metal powders used in laser sintering systems and electron beam melting, are just such a case. By way of an example, a UK-based AM company called Metalysis, commercialized a process for refining and preparing titanium powder directly from titanium ore (called FFC). The process is one that requires less energy to produce titanium powder than the current established process. Moreover, the process employs a non-toxic chemical, calcium chloride, during refinement and any leftover amount can be reused. Nevertheless, considering the immaturity of the technology there are still a few materials for which such novel processing techniques exist.

As mentioned before, the environmental impacts and benefits of AM technology can be obtained from various phases including component and product redesign, overall process redesign, material inputs, production and related strategies, and by "closing the loop". Table 3.3 details the sustainability benefits throughout the product lifecycle. These advantages have been realized in the industrial companies listed in the table. These companies are clustered based on the lifecycle stages on which they focus. For instance, Rolls-Royce benefited from using AM to redesign its components, resulting in environmental improvements in the production phase. They were able to reduce energy intensity and waste material in production, in addition to extending the product lifecycle through in situ services for maintenance and repair (Ford and Despeisse 2016).

Exhibit 1—Environmental Impact—AM Technology as a Driver for Sustainability

AM has many impacts on the environment as discussed in this chapter. In terms of energy consumption, studies demonstrated that the energy-related efficiency of AM is not better than that of conventional manufacturing methods such as injection molding, except in some situations (Khorram Niaki and Nonino 2017a). For instance, AM has an advantage in terms of energy considerations, when the production volume is small. Moreover, the prospect of design optimization offered by AM may also result in a reduced energy demand. Furthermore, since production-related energy is linked to production time, production time is linked to product height (number of layers), thus it is expected that conventional manufacturing will a have much better performance for a larger product size than AM.

In contrast, considering a product's lifecycle from raw material to logistics and use phase, AM may have profound impacts on the environment. For instance, it can reduce fuel consumption through reducing transportation, and it certainly reduces waste and material usage.

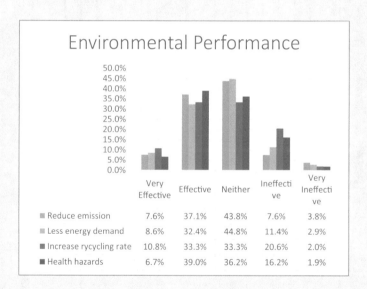

	Very Effective	Effective	Neither	Ineffective	Very Ineffective
Reduce emission	7.6%	37.1%	43.8%	7.6%	3.8%
Less energy demand	8.6%	32.4%	44.8%	11.4%	2.9%
Increase rycycling rate	10.8%	33.3%	33.3%	20.6%	2.0%
Health hazards	6.7%	39.0%	36.2%	16.2%	1.9%

Considering the results of the survey, implementing AM technologies may have slightly environmental benefits. Among which, more than 45% perceived the effectiveness of AM in eliminating health hazards, less than 45% for reducing emissions, 44% for increasing material recycling rate, and 41% for reducing energy demand. However, as shown in Chap. 6, it depends on the type of industry, raw material and the AM technologies in question.

Table 3.3 AM sustainability benefits across various lifecycle stages as seen in industrial examples

Section	Example	Product/process design	Material processing	Component/product manufacturing	Use/service	Repair/re-manufacturing	Recycling
Component and product redesign	SAVING project			✓	✓		
	Rolls-Royce			✓	✓	✓	
	GE			✓	✓		
Process redesign	Salcomp			✓			
	Construction			✓			
Material input processing	Metalysis		✓				
	Filabot		✓				✓
	EKOCYCLE Cube		✓				✓
	Bewell Watches		✓				✓
Make-to-order component and product manufacturing	Kazzata	✓				✓	
	Siemens			✓		✓	
	Home 3D printers	✓		✓			
	3D Hubs			✓			
Closing the loop	PPP		✓				
	Caterpillar				✓	✓	✓
	HMT			✓		✓	

Source Ford and Despeisse (2016)

3.3   Societal Sustainability

The social implications of AM remain poorly studied at the time of writing and few studies exist on this topic. Nevertheless, it is possible to make some important considerations as the advent of this emerging manufacturing paradigm may also have other major social implications in the areas of geopolitics, employment, demographics, patterns of consumption and security.

In terms of employment implications, AM has different impacts on developed and developing countries, as in a developed country with an (relatively) ageing population, the lack of an available workforce exist, so AM—in a similar way to other hi-tech methods—can be considered as beneficial tools for competitiveness. In addition, it causes decreasing operational costs through a lower demand for labor, particularly in developed countries where the average cost of labor is significantly higher than in developing countries. However, the reduced labor demands for developing economies are expected to contribute to socioeconomic instability. In contrast, AM offers many opportunities to remote areas with poor economic situations. The simple process of 3D printing and less resource requirements enables less-developed regions to bridge the gap to high-tech manufacturing methods and to join this new emerging market along with advanced countries. It supplies these areas with the objects required for improving the quality of life, and it also reduces the need for imports from advanced areas and promotes independence, since they are now able to design and produce on-site whatever they need.

A change in the labor market is one of the most important social implications of the use of additive manufacturing. Following Chen et al. (2015), some social dimensions and indicators, which can help our understanding of the situation are as follows:

- Working condition: number of work injuries per year and the costs of injuries caused by machine or dangerous toxic gas
- Work's impact on the long-term health of workers: number of long-term injuries per year, such as depression, back pain and lung cancer caused by working conditions
- Employee turnover: the rate expressed as a percentage at which the factory gains and loses employees, since it reduces companies' productivity if they have too many employees.
- Proportion of permanent employees: the ratio [percentage] of permanent to non-subcontract employees. This number is an important corporate responsibility factor, indicating the factory's commitment towards its workers.
- Employee empowerment: the number of training seminars the company offers to improve their employees' ability to complete their designated tasks.

AM may have health benefits for workforces when compared to conventional manufacturing processes as it allows workers to avoid a long-term presence in harsh and potentially hazardous working environments. As little research has been conducted on the toxicity of materials and AM processes, it is possible that such impacts may exist during those process and in the disposal of any materials used. One study on social issues focuses on working conditions and worker health as social indicators (Huang et al. 2013).

Table 3.4 reports the occupational hazards and environmental impacts of different AM processes such as SLA, SLS, and FDM and the information is based on the chemicals used in the system. As seen in the table, most of the materials are not harmful to humans, except for the photopolymers and propylene glycol. However, workers and operators must be informed of the potential safety hazards, especially in the handling and disposal of these chemicals. Obviously, some safety equipment such as masks, goggles and gloves should be provided for the operators.

As will be reported in Chap. 4, the introduction of additive technologies to manufacturing and business has shifted the current business and manufacturing paradigms, leading societies to new entrepreneurial opportunities.

Supply chains are being de-globalized with the production of orders becoming on-demand and closer to consumers. The localization of manufacturing could have potential opportunities for the reduction of global economic imbalances. Therefore, though the increasing spread of AM, a profound economic impact on countries' reliance on import/export is expected. Many countries whose prosperity and political power are linked to their exports, especially of consumer products, such as China and Japan might be severely affected as demand for imports falls. The shift in global manufacturing to AM technologies is expected to reach trillions of dollars in business over the coming decades (Campbell et al. 2011). This business volume involves 3D printed objects, 3D printers and materials, professional services including designing and legal fees earned in intellectual property (IP) protection and dispute settlements.

The advent of this emerging technology is promoting a new digitalized business model and supply chain, and as a result more information technology-related skills are expected in order to develop the new business chains and serve its functions. The emerging business areas include the market for selling/purchasing 3D design files, the existing open-source market for assisting 3D production (theses aspects of AM technologies will be discussed in Chaps. 4 and 5). AM also creates the need for training and education in digital manufacturing, especially when the manufacturing environments encounter a shift in design thinking. Therefore, educational programs should be adjusted to meet the new knowledge and skill demands. This can further be considered as an opportunity to compensate for the elimination of manual work in manufacturing.

Furthermore, new regulatory frameworks should be adjusted to this new manufacturing and business areas, in which the digitalization of resources, products, designs and ideas has been more achievable. This framework should involve clear legal rules concerning digital blueprints of objects, which used to refer to a reproduction of a technical drawing or an engineering design using a contact print process on light-sensitive sheets. This should also involve other aspects of the business such as 3D scanning or the online distribution of ideas. Licensing, copyright and patent patterns might be significantly affected since AM enables an easy and simple way by which to reproduce a part, particularly in combination with

Table 3.4 Environmental impacts and occupational hazards of different chemicals in AM processes

AM process	Chemical/solvent	Emissions	Occupational hazards	Biogradability
SLA	Propylene carbonate	CO_2, CO, SO_x	Low system toxicity	Readily biodegradable
	Urethane resins		Too much ingestion may lead to vomiting	Not dangerous to the environment
	Tripropylene glycol		Slight irritation after eye contact	Biodegraded (50% in 8.7 days, and by 81.9% over a 28-day)
	Isopropanol		Eye irritation and burning and sometimes corneal injuries; Skin irritation and soreness; prolonged exposure may cause dermatitis	Potential to highly decline oxygen from aqueous systems
SLS	Polyamide resin	CO_2	No serious hazards	Inflammable or long exposure to air
	Acrylonitrile butadiene styrene		Molten plastic likely to cause lethal burns, processing fumes may lead to eye irritation and choking of the respiratory tract	Since it is insoluble in water, its eco-toxicity is low
FDM	Propylene glycol monomethylether	CO_2, CO, SO_x, PMc, NO_x	Irritation in eyes, skin, nose, throat; headache, nausea, dizziness, drowsiness, incoordination; vomiting, diarrhea	No hazardous decomposition products

Source Huang et al. (2013)

3D scanning. Therefore, regulatory frameworks are needed to encompass all aspects of digitalized designs and manufacturing. This should include limitations and regulations on controlling possible harmful outcomes.

Another social dimension deals with the capability of AM in democratizing the designing and production of goods, and usually refers to direct digital manufacturing. Thanks to the existing information technologies, widely available 3D printers and CAD software, patterns of consumption have been subjected to a big shift. End users are now enable to make their products themselves, rather than being passive consumers. Instead, they are becoming prosumers in a global manufacturing community (Chen et al. 2015). Online 3D printing platforms (detail in Sects. 4.1.3 and 4.3) are expected to provide various opportunities to private users (Anderson 2012). Many spare parts, food or lab equipment can now be 3D printed on-demand at home. Therefore, it is expected that the majority of prosumers are unaware of the environmental implications of AM, and as a consequence sustainability impacts may not be considered by them (Kohtala and Hyysalo 2015).

Furthermore, online 3D printing platforms provide an easy means of producing everything we want, even though it can also be a security threat at the same time. For instance, the blueprints of weapon designs may result in the private fabrication of firearms or other weapons, over which governments have no control, resulting in illegal behaviors and violations of the law (Simon 2013).

Generally, in terms of security implications, AM technology may have significant implications as follows (Campbell et al. 2011):

– Weapons fabrication such as guns, bullets, bombs, etc., will become easier, cheaper and more easily accessible
– Weapons can be much more easily disguised (e.g., IEDs (improvised explosive devices) that look identical to non-weapons)
– Terrorists can produce to their needs, being independent of their suppliers in advanced countries, for instance
– Implications will exist for counterfeiting/anti-counterfeiting.

Seven Key Facts

• AM will result in cost reductions of 170–593 billion USD, an avoided Total Primary Energy Supply (TPES) of 2.54–9.30 exajoules (EJ) and avoided CO_2 emissions of 130.5–525.5 Metric tons (Mt).
• AM technologies are a novel and advanced Flexible Manufacturing System.
• AM technology will drive the shifting of production locations nearer to, or even at, the point of end consumers, resulting in geographically delocalized production.
• Studies demonstrate the efficiency of conventional manufacturing rather than additive process for large production volumes: AM is a very competitive alternative for smaller production volumes.
• AM can enable the reduction of energy consumption through operations outside of manufacturing processes such as reducing or eliminating transportation, and inside manufacturing processes through reducing material waste.
• Considering the increasing level of AM spread, a profound economic impact on countries' reliance on import/export is expected.
• Licensing, copyright, and patent patterns might be significantly affected, since AM provides an easy means of reproduce parts.

References

Anderson, C. (2012). *Makers—The new industrial revolution*. New York/USA: Crown Business.

Baumers, M., Tuck, C., Wildman, R., Ashcroft, I., Rosamond, E., & Hague, R. (2013). Transparency built-in. *Journal of Industrial Ecology, 17*(3), 418–431.

Bourell, D. L., Leu, M. C., & Rosen, D. W. (2009). Roadmap for additive manufacturing: Identifying the future of freeform processing. The University of Texas at Austin, Austin, TX, 11–15.

Campbell, T., Williams, C., Ivanova, O., & Garrett, B. (2011). *Could 3D printing change the world? Technologies, potential, and implications of additive manufacturing*. Washington, DC: Atlantic Council.

Chen, D., Heyer, S., Ibbotson, S., Salonitis, K., Steingrímsson, J. G., & Thiede, S. (2015). Direct digital manufacturing: definition, evolution, and sustainability implications. *Journal of Cleaner Production, 107*, 615–625.

Cozmei, C., & Caloian, F. (2012). Additive manufacturing flickering at the beginning of existence. *Procedia Economics and Finance, 3*, 457–462.

Drizo, A., & Pegna, J. (2006). Environmental impacts of rapid prototyping: an overview of research to date. *Rapid Prototyping Journal, 12(2)*, 64–71.

Elkington, J. (2002). *The triple bottom line*. London: Capstone.

Faludi, J., Bayley, C., Bhogal, S., & Iribarne, M. (2015). Comparing environmental impacts of additive manufacturing vs traditional machining via life-cycle assessment. *Rapid Prototyping Journal, 21*(1), 14–33.

Ford, S., & Despeisse, M. (2016). Additive manufacturing and sustainability: An exploratory study of the advantages and challenges. *Journal of Cleaner Production, 137*, 1573–1587.

Gebler, M., Uiterkamp, A. J. S., & Visser, C. (2014). A global sustainability perspective on 3D printing technologies. *Energy Policy, 74*, 158–167.

Huang, S. H., Liu, P., Mokasdar, A., & Hou, L. (2013). Additive manufacturing and its societal impact: A literature review. *The International Journal of Advanced Manufacturing Technology, 67(5–8)*, 1191–1203.

Khorram Niaki, M., & Nonino, F. (2017a). Additive manufacturing management: A review and future research agenda. *International Journal of Production Research, 55*(5), 1419–1439.

Khorram Niaki, M., & Nonino, F. (2017b). Impact of additive manufacturing on business competitiveness: A multiple case study. *Journal of Manufacturing Technology Management, 28*(1), 56–74.

Kohtala, C., & Hyysalo, S. (2015). Anticipated environmental sustainability of personal fabrication. *Journal of Cleaner Production, 99*, 333–344.

Le Bourhis, F., Kerbrat, O., Hascoët, J. Y., & Mognol, P. (2013). Sustainable manufacturing: Evaluation and modeling of environmental impacts in additive manufacturing. *The International Journal of Advanced Manufacturing Technology, 69(9–12)*, 1927–1939.

Lozano, R. (2008). Envisioning sustainability three-dimensionally. *Journal of cleaner production, 16*(17), 1838–1846.

Luo, Y., Ji, Z., Leu, M. C., & Caudill, R. (1999). Environmental performance analysis of solid freedom fabrication processes. In Electronics and the Environment. *Proceedings of the 1999 IEEE International Symposium* (pp. 1–6). IEEE.

McKinsey Global Institute. (2013). http://www.mckinsey.com/.

Sharif Ullah, A. M. M., Hashimoto, H., Kubo, A., & Tamaki, J. I. (2013). Sustainability analysis of rapid prototyping: Material/resource and process perspectives. *International Journal of Sustainable Manufacturing, 3(1)*, 20-36.

Simon, M. (2013). When copyright can kill: How 3D printers are breaking the barriers between intellectual property and the physical world. *Pace Intellectual Property, Sports and Entertainment Law Forum, 3*, 60.

Song, R., & Telenko, C. (2017). Material and energy loss due to human and machine error in commercial FDM printers. *Journal of Cleaner Production, 148*, 895–904.

Weller, C., Kleer, R., & Piller, F. T. (2015). Economic implications of 3D printing: Market structure models in light of additive manufacturing revisited. *International Journal of Production Economics, 164*, 43–56.

Yoon, H. S., Lee, J. Y., Kim, H. S., Kim, M. S., Kim, E. S., Shin, Y. J. … Ahn, S. H. (2014). A comparison of energy consumption in bulk forming, subtractive, and additive processes: Review and Case study. *International Journal of Precision Engineering and Manufacturing-Green Technology, 1(3)*, 261–279.

Chapter 4
The Value for Business and Operations Strategy

This chapter begins with a discussion of the value of AM for business strategies, and describes the impacts it has on a firm's competitiveness. It then shows how AM can drive creativity and innovation, and the nature of the innovation that may occur thanks to AM. Then, the capability of the technology to offer new online services is explained.

This chapter also outlines some of the likely key changes and their effects on operations strategy and consequently on manufacturing and supply chain management paradigms. In fact, it provides answers to two main questions: "*How will AM change operations and supply chain management paradigms?*", and "*How will AM change the manufacturing supply chain process and components?*" The role of AM technologies in manufacturing and supply chain paradigms will also be analyzed. In the latter case, these supply chain paradigms include several concepts such as leanness, agility and mass customization: AM offers the real leagility features through enabling responsive manner for innovative products. AM also preserves the features of a lean supply chain (meaning there is less waste), and the agile paradigm (flexibility and time). We discuss the impacts that adopting AM has on operations using empirical evidence, and we evaluate potential approaches. The value for customers and the mass customization paradigm are then discussed, through which the benefits for entrepreneurial opportunities and prosumers are also discussed. The chapter ends with the strategic and technological barriers facing the spread of AM.

4.1 The Value for Business Strategy

4.1.1 AM as a Driver for Business Competitiveness

The numerous advantages of AM can enhance the competitiveness of both a company and a product. Companies from developed countries have been encountering the challenge of losing competitiveness to mass production. Therefore, AM

© Springer International Publishing AG 2018
M. Khorram Niaki and F. Nonino, *The Management of Additive Manufacturing*,
Springer Series in Advanced Manufacturing,
https://doi.org/10.1007/978-3-319-56309-1_4

can be seen as a powerful tool that offers the necessary competitiveness, by producing high-value and low-volume complex parts. Khorram Niaki and Nonio (2017b) using multiple case study approach demonstrated that SMEs adopting AM for producing end-usable products could increase their competitiveness.

Moreover, each industry has its own competitive requirements. These requirements can be either design-based, production-based, or processing-based, or can be a combination of all three. For instance, in several application sectors such as aerospace, automotive and industrial components, efficient operations, along with weight and material cost savings are of the utmost importance, while for healthcare industries it is increasing functionalities and the integration of components, and for consumer goods it is the degree of customization that are the most important factors (uz Zaman et al. 2017).

Exhibit 2—Impact on Business Competitiveness—Catching New Opportunities
A fast response to business opportunities and flexibility in manufacturing can be key factors in determining the competitiveness of a company. It can better satisfy the requirements of different customers and it therefore provides products with greater global competitiveness. It may provide a faster response to business opportunities by fully introducing the technology to its new product development process. Lower operational costs can also be considered as a competitive advantage, and additionally, an innovative product—achievable through the use of AM—can also increase the ability of a company to compete.

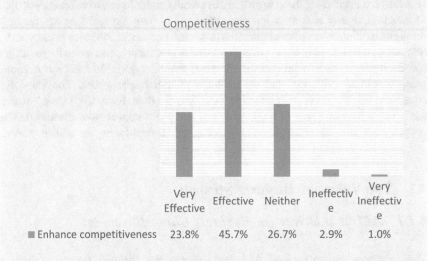

	Very Effective	Effective	Neither	Ineffective	Very Ineffective
Enhance competitiveness	23.8%	45.7%	26.7%	2.9%	1.0%

The results of our survey demonstrate that more than 69% of the companies questioned affirm that AM has the potential to increase business competitiveness.

Exhibit 3—Impact on Business Competitiveness—Accessing New Markets and Acquiring New Customers

AM technologies are most advantageous in market environments characterized by a demand for customization that can increase market share through acquisition of a broader customer domain (i.e. Nike launched a customizable football cleat produced using AM in 2013). The acquisition of new customers refers to the increasing demands in the same market. These two tactics are offered by AM through its various capabilities such as the provision of low cost modification of designs and an efficient method for customization, removing the constraints of conventional manufacturing on both design and production, and enabling toolless manufacturing (further explanations of these aspects are provided in Chap. 5). These impacts facilitate a faster response to customer needs, resulting in the acquisition of new customers.

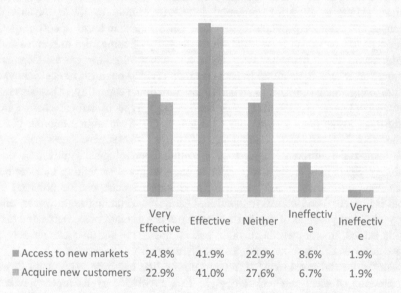

New markets and new customers

	Very Effective	Effective	Neither	Ineffective	Very Ineffective
■ Access to new markets	24.8%	41.9%	22.9%	8.6%	1.9%
■ Acquire new customers	22.9%	41.0%	27.6%	6.7%	1.9%

As reported above, more than 66% of the companies surveyed affirm that AM is helpful in gaining accessing to new markets, and about 64% affirm that it is helpful in acquiring new customers in existing markets.

4.1.2 AM as a Driver for Creativity and Innovation

Improving the creativity and innovativeness of companies and individuals is a major goal of AM technologies, as they provide the capability of producing almost everything that one can draw. It is more beneficial than conventional manufacturing, where the first version of a product is only achievable by following an expensive and slow process, which could result in the loss of several people's creative ideas or the missing of opportunities due to slow product launching. In order to understand AM's implications on innovation, we first need to know which types of innovation are offered by AM technologies.

Looking at the literature on technological innovation, various phrases and classifications can be found. These include incremental innovation vs radical innovation; disruptive innovation and technology; and revolutionary technology. To understand the position of these emerging AM technologies, a clear definition of each one is needed.

For a long time, innovation has been seen as either incremental or radical in nature (Marquis 1969). Incremental innovation refers to those technological changes that improve products' or services' performance in terms of cost or quality. This type of innovation is essential for the majority of companies to survive and be competitive, and it happens every day in a company. In other words, incremental innovation results in minimal impacts on industry. By contrast, radical innovation refers to big changes, providing a brand-new functional capability. It happens rarely and discontinuously, resulting in a breakthrough type of radical change in an industry. Furthermore, Bower and Christensen (1995) identified another classification that they termed sustaining and disruptive technologies. These can help to distinguish the different impacts of a certain technological innovation on an industry. Sustaining technologies refer to one that gives something more or better than what customers already receive. These improvements can be obtained from both incremental and radical innovation. Disruptive technologies, however, offer a very different package of attributes from what mainstream customers have historically valued. In other words, disruptive technologies change the game entirely, and the current established technologies are disrupted from the market.

An "industrial revolution" is another phrase that has been used many times in discussions of emerging technology. In fact, revolutionary technology refers to something that fundamentally shifts the way things happen. These shifts do not include only growth in productivity or other technical advances, they also relate to fundamental shifts, for example in what was done previously in manufacturing. In fact, it refers to the technology that has widespread and structural impacts in society. During the last few centuries, three industrial revolutions were identified. The first industrial revolution occurred in UK from 1750 to 1815, and involved technologies related to steam power. The second was from 1870 to 1914 in USA and Germany and involved technologies related primarily to the development of chemicals and electricity. The third industrial revolution was from 1973, and involved technologies related primarily to advancements in information technologies and communication. It occurred in USA and East Asia. Technological innovation, whether incremental or radical, may result in a wide range of changes to markets, some very disruptive and some not.

The text above provides insights into the types of technological changes and the level of their impacts. Several researchers employed this classification to introduce AM technologies. The question is, if it is an incremental, radical, disruptive or even industrial revolution. More recently, Steenhuis and Pretorius (2017) studied these issues.

Obviously in terms of incremental innovation, AM technologies have found numerous successful applications, for instance in reducing costs, time to market and reducing the amount of waste, and these are mentioned in Chap. 2 as they relate to different industries such as the aerospace or automotive industries. It was shown that AM enables incremental improvements when compared to traditional manufacturing.

As for radical innovation, AM has been most successful in the medical industry, for instance in the production of prostheses or surgery guides that were conventionally very expensive or impossible to produce. It enables producers, even without medical knowledge, to fabricate a medical component rapidly and at a

Fig. 4.1 Prosthetic hand. *Image source* Student's facebook page

dramatically lower cost. According to Love (2014), a group of students with a consumer-level 3D printer (Makerbot) could fabricate a prosthetic hand for just $10 (Fig. 4.1). The manufactured prosthesis could then be used by a girl who was born without fingers. Nevertheless, using conventional manufacturing to create the same part used to cost more than $50,000. In this case, the designed file can be downloaded from available websites, and in just a short time the design can be 3D printed. This example has been about customization at newly affordable costs to provide unique solutions that were not possible before.

These websites are examples of "Enabling the Future", and provide useful information on the functionality of each design. One more example is the case of surgery planning that was explained in Chap. 2. Here, the surgeon can have and touch the mockups pre-operation, which used to very expensive or in many cases even impossible. The application of AM for in situ manufacturing, for instance additively producing houses with a concrete material will have huge advantages and could lead to radical changes in the construction industry. A schematic of this process is shown in Fig. 4.2. However, in this industry, AM still has a relatively long way to go before it is at a widespread level. In addition, the capability of AM to make highly complex parts (see example in Fig. 4.3) which were not possible with conventional manufacturing—or at least were not at the same quality—can be considered as radical changes in manufacturing. Therefore, these examples show how AM technology can offer radical innovation, and breakthrough technologies that can radically affect an industry.

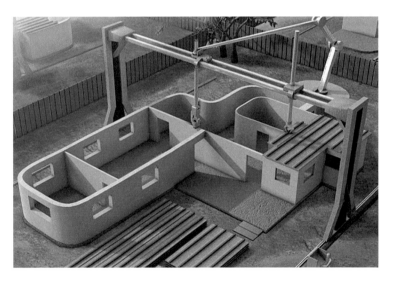

Fig. 4.2 A schematic of an in situ 3D printing of a house. *Image Source* Flicker

Fig. 4.3 An example of a complex 3D printed part. *Image Source* Flicker

The technology was first adopted with the aim of using it for prototyping and this still remains the most widespread application. It provides a very fast and cost-effective process for making prototypes or manufacturing tools. In this case, it serves the existing customers. Therefore, these improvements depend upon the actual impacts, offering incremental and radical innovation, and not disruptive technology. However, it can be considered as a disruptive technology when it is being adopted by manufacturing companies that have relied on conventional manufacturing methods such as the injection model. However, as mentioned in Chap. 3 in the economic section, AM can currently be considered as a competitive method for products of a small size and with a medium production volume. The rapid technological advancements in recent year promise a workable and economic manufacturing method for the production of larger volumes in near future. Therefore, as for disruptive technology, AM has a major potential to be a disruptive technology, specifically when it becomes a competitive alternative to conventional manufacturing methods.

By the time that AM becomes a disruptive technology for conventional manufacturing, most organizations will find it difficult to compete and adapt to it. Therefore, organizations must address the fundamental concern regarding management constraints by answering the following question: *what has to change in my business in order for it to be successful?* This will help policymakers and strategic managers to boost the usefulness of the technology. They must employ a new business model based on the characteristics of the technology and its potential opportunities.

An "industrial revolution" is supposed to have fundamental changes not only to manufacturing but also in society. These types of impact were seen in the first industrial revolution, where numerous shifts occurred such as the move to industrial manufacturing instead of focusing on small-scale production or being more agricultural, moving from villages to urban environments, and increasing the need for education. As discussed in Chap. 3, the social implications of AM have brought and will bring more widespread changes. 3D printing technologies hold the promise to change the way we consume, create, and maybe even live in the world. New education and training systems are needed to prepare people to realize AM's capabilities and potential opportunities. For instance, many elementary and high schools that provide 3D printing to educate students in order to initiate the education for them at the beginning of learning process, currently adopted to curriculum in USA and many other developed countries.

The change in education is mostly aimed at developing the culture of the new manufacturing method, enabling shifts in manufacturing from big factories and long production lines to home fabrication. It provides space in universities and libraries for people to physically produce from their imaginations and to their needs. Moreover, AM enables an economy-of-one that means the possibility exists of producing a single customized or complex design in a cost-effective manner and in many cases at an even cheaper cost than that which the manufacturers offer. It makes real the personal manufacturing by consumers that is a major challenge to industrial manufacturers. As mentioned before it also has geopolitical effects since it enables less developed countries to compete without owning current high-tech equipment. It also provides novel business models especially for consumer products produced in houses. However, these social changes are currently not big enough to be considered as a new industrial revolution. In fact, at the time of writing it is too early to say that AM technologies will be the initiator of the next industrial revolution.

> **Exhibit 4—Impact on the New Product Development Process—AM Support Innovation and Creativity**
> Is AM a successful technology for increasing creativity and innovativeness? Yes, AM brings innovativeness through different factors. These factors include removing conventional production constraints and enabling freedom of design and producing parts which are geometrically more complex. It is also capable of producing integrated parts that do not need to be assembled, simplifying design modification and iteration, easing up the production process, and dematerializing the supply chain (Khorram Niaki and Nonino 2017a). Moreover, it can achieve a higher strength-to-weight ratio by enabling the production of lightweight parts and through the use of alternative raw materials. A detailed discussion on innovation through AM is provided in Sect. 4.1.2. Thus, AM has brought to the fore many possibilities for increasing innovation such as offering new collaborative forms of design and manufacturing, and using crowd-based innovation and democratizing design.

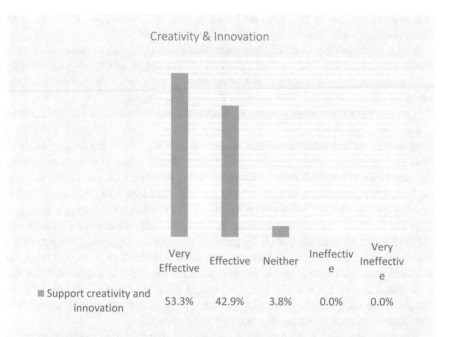

The impact of AM on business flexibility is widely recognized. An interesting result is that nobody believes that AM does not assist innovativeness, and only three out of 105 companies says that AM is unsuitable for customization. As regards creativity and innovation, more than 96% perceived the effectiveness of AM in assisting innovativeness. This is due to the flexibility of AM that brings in all processes of product development from the making of early conceptual models, through functional tests before production and design modification, to the final production stage. It can fabricate almost anything that can be designed.

4.1.3   AM as a Driver for New Services: The Online 3D Printing Platforms

Online 3D printing services and platforms have increasingly appeared during recent years. These platforms seek to serve the special needs of their customers. Currently, in consumer communities there are relatively few 3DP owners. In addition, those 3DP owners can only fabricate parts made of plastic materials, or the machine does not meet their requirements (e.g. in terms of size, mechanical properties, and color, etc.). Some other consumers may also need a design of a specific product so that they can print it by themselves at home. Therefore, consumer communities have

two main needs, firstly for those that need to print their designed object and secondly for those who need a design file to be 3D printed.

The available online AM services serve these two main needs; some work only
for the manufacturing of the object designed by consumers, while some others offer
only designed files, and others still cover both requirements as the detail of the
business model of online 3DP services illustrates in Fig. 4.4.

The designed files sometimes are accessible free and in some case is provided for
a fee. These platforms also host designs from third party designers, creating a
marketplace for individual designers. These designs are sold in the host system and
both designers and hosters benefits from their services. There is also an option for
the consumer to customize the available designs, whether their own or third party
designs. This customization may include object features such as size, shape, color,
or body fitting. These service systems can also assist consumers during the design.
This can include various services such as converting two-dimensional sketches to
3D model data that is then usable by printing machines, transferring pictures to the
usable file, or even creating 3D model data from a physical object by using a 3D
scanning process. These services also provide for collaborative designs or the use of
crowdsourcing, where one can order the design from the "crowd" and explain the
details of the required object and its characteristics, so that the available designers
will be able to design the product exactly as required.

For manufacturing purposes, these service systems facilitate the production of
the designs, which are then shipped to the customer or delivered to the store. In
addition, they supply home-use 3D printers to be owned and used by consumers at
their home or at their office. They can also work as an intermediary between those

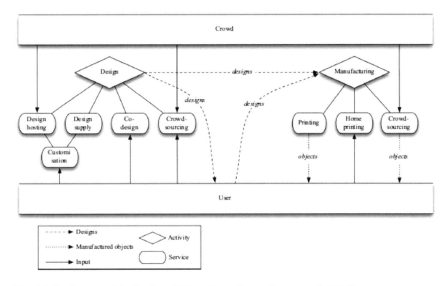

Fig. 4.4 Business model of online 3DP services. *Source Rayna et al.* (2015)

users that want to have a 3D printed part of a design and those that own 3D printers and want to do the work for a fee.

Therefore, the main types of service offered by 3D printing platforms are as follows:

- Supplying designs
- Hosting designs
- Customization of designs
- Co-designing
- Crowdsourcing designs
- 3D printing of objects
- Selling 3D printers
- Crowdsourcing 3D printing.

Table 4.1 illustrates leading 3D printing service systems, outlining which services they provide. These platforms are categorized in terms of design marketplace, 3D printing services, 3D printing marketplace, and crowdsourcing platform. The table is from the study of Rayna et al. (2015).

As seen in the table, the majority of the companies provide a marketplace for designers (third party) to host and sell their designs. However, a few of them supply their own designs as well (Sculpteo, and i.Materialise). Several companies (e.g. 3DaGoGo, Shapeways, 3DCreation Lab) also assist users who want to transform their idea to the final 3D model data, through co-designing with them. In addition, some others provide design crowdsourcing. This service enables individuals to engage the help of the online community in the creation of their designs. The design obtained from the crowd (or the user's own design) can become a physical object through different pathways:

- 3D printing at home by users through purchasing 3DP from machine suppliers (e.g. iMakr, Makerbot),
- Paying for 3D printing services online (e.g. Sculpteo, and Materialise Onsite),
- Fabricating through a physical store (e.g. iMakr, and WhiteClouds),
- Making the object by crowdsource 3DP equipment (e.g., 3D Hubs, and MakeXYZ).

4.2   The Value for Operations Strategy

A business strategy guides the so-called operations strategy. Traditionally, an operations strategy concerns the total transformation process, namely the whole business and its objectives regarding the improvement of external performances (those evaluated by customers) of cost, time, quality and flexibility. The operations strategies of companies differ because companies choose different performance objectives, which become their competitive priorities for gaining a competitive advantage. The decisions taken as part of a company's operations strategy are

Table 4.1 Online 3D printing platforms

Platforms	Headquartered in	Design						Manufacturing			
		Hosting (sales)	Hosting (repository)	Design supply	Co-design service	Design crowdsourcing	Design customization	3DP service	3DP sales	Physical store	3DP crowdsourcing
3DaGoGo	USA	*									
3D Burrito	Sweden	*									
Cubify cloud (acquired by Microsoft's 3D builder in 2015)	USA	*		*			*	*	*		
i.Materialise	Belgium	*		*	*		*	*			
Ponoko	New Zealand	*						*			
Sculpteo	France	*		*	*		*	*			
Shapeways	Netherlands	*			*			*			
Trinckle 3D	Germany	*				*		*			
3DPrintUK	United Kingdom							*			
Materialise Onsite	Belgium							*			
3DCreation Lab	United Kingdom				*			*			
FastProtos.com	USA				*			*			
iMakr	United Kingdom				*		*	*	*	*	
Makerbot/Thingiverse	USA		*						*	*	
The 3D Printer Experience	USA				*			*		*	
WhiteClouds	USA			*	*			*		*	
3D Hubs	Netherlands										*
MakeXYZ	USA										*
Additer	Australia					*					*
Kraftwurx	USA					*					*
Maker6	Canada	*			*			*			*

Adapted from Rayna et al. (2015)

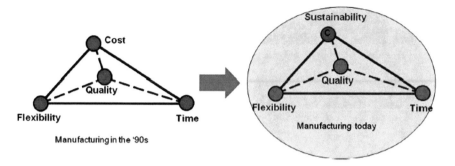

Fig. 4.5 Manufacturing attributes in the 1990s and at the present time. *Source* Salonitis and Stavropoulos (2013)

strategic, because they affect the long-term development of operations resources and processes and are the basis for a sustainable advantage (Slack and Lewis 2003).

Some companies currently consider the impact of operations on energy consumption, and the efficient use of resources and societal aspects. These aspects are related to sustainability. Therefore, it is evident that the sustainability factor incorporates the cost, as seen in Fig. 4.5—the shift from cost to sustainability. The role of AM in driving sustainability was discussed in Chap. 3. Here we present the value of AM for a business and operations strategy.

4.2.1 Impact on Manufacturing Paradigms

The introduction of additive technologies to manufacturing and business has shifted the current manufacturing paradigms. These paradigms can be classified in four different typologies (Fig. 4.6):

- Craft manufacturing
- Mass manufacturing
- Mass customization
- Direct digital manufacturing.

Craft production involves experts who rely on their own skill and knowledge for specialized tasks and product design. There are few consumers for these type of products, which mostly involve the local community and are usually at small production volumes. Mass production instead relies on more standardized products and production in a large factory with large production volumes based upon specific designs. In the mass customization paradigm, the production volume is the same as that in mass production although customers have choices in product design. The products are usually modularized or bespoke for specific groups of end users. Ultimately, what AM technologies—as direct digital manufacturing methods—will add to these paradigms is the direct choice of consumers with respect to product

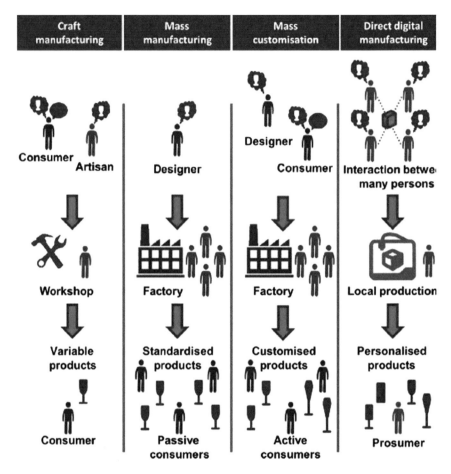

Fig. 4.6 A comparative schematic of manufacturing paradigms and their main actors. *Source* Chen et al. (2015)

designs. It may include the prosumers (producers that are also consumers of the product); customized or individualized products; localized production and with most, automated manufacturing processes. The details of entrepreneurship and business opportunities for individual users are provided in Sect. 4.3.2.

From a company's point of view, AM exclusively empowers manufacturers to produce fully customized products in a sustainable manner. According to the Cambridge online dictionary, "customize" means: *"To make or change something according to the buyer's or user's needs."* AM allows designers an almost unlimited freedom of design, resulting in real mass-customization of consumer goods. In addition, through the elimination of molds and any extra tools, the process of customization becomes technically feasible and cost effective. Thus, in comparison with existing methods of product customization, AM reduces waste, costs and the number of necessary manual operations.

Exhibit 5—Impact on a New Product Development Process—AM as a Driver For Customization
Customization through conventional manufacturing is neither technically possible, nor cost effective since it requires many changes to production lines, molds, machining and tools in order to fulfill customer expectations, resulting in extra costs. In contrast, AM has overcome these two limitations. AM has been an effective method for product customization.

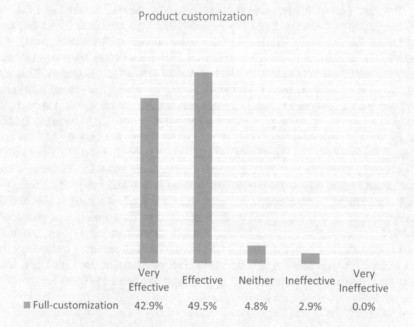

More than 92% of respondents affirmed the effectiveness of AM for customization. The results demonstrate the fact that the capability of AM for customization is not related to specific types of industry, operation, or product. Consequently, all industries and application sectors are encouraged to implement AM technologies to develop their business, and by doing so they may acquire a broader range of customers and markets.

In the investigation of the impact of AM on operation strategies, the paradigm of mass customization must be taken into account. Mass customization has substantial impacts on the production environment and could be thought as a key driver of the agile supply chain's prominence in manufacturing and business thinking world-wide. The use of mass customization can be found in a variety of industries, for instance in the electronics industry for PCs and cell phones.

Alford et al. (2000) studied the concept of mass customization in the vehicle manufacturing industry. They defined three strategies for customization in the automotive industry. These include core, optional and form customization. Core customization refers to direct customer involvement in the design process that requires a low-volume product. Optional customization, by way of contrast, focuses on the high-volume products where the customer chooses a customized option. Form customizations refers to those that need to be changed from the standard form, for instance, these may include changing the finance package, service and warranty specifications.

The mass customization method generally uses a standard product as a base for customization and has the flexibility of individual customization. Given that core customization is already impossible with conventional methods (Fox 2003), AM enables the production of complex parts from digital data taken from the customer, meaning that there is a customer involvement in product development. This idea was applied to body-fit parts (Tuck et al. 2008), in other words customized seats and seat backs, in order to provide a more comfortable environment. Although, the body-fit core customization process is not exclusive or novel, AM certainly enables costly benefit customization. Tuck et al. (2008) found the role of AM on conventional manufacturing paradigms. Manufacturing paradigms differ from low to high variety products and their respective production volumes. Traditional customization occurs where the production volume is low and the product variety is high. In addition, mass production occurs when the product variety is low and the production volume is high. This conflict between product variety and production volume causes the emergence of the flexible manufacturing system and mass customization. However, the advent of AM as a new manufacturing technology that is capable of high product variety without the requirement for intensive labor, leads to a possible alternative manufacturing paradigm as shown in Fig. 4.7.

Fig. 4.7 Manufacturing supply characteristics, including AM. Adapted from Tuck et al. (2008)

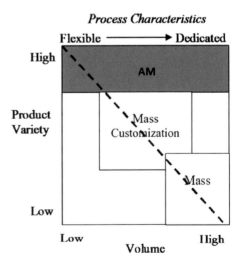

4.2.2 Impact on Product Lifecycle and Operational Costs

Among several potential sustainability impacts of AM, the following two benefits can be considered as significant opportunities as stated by Ford and Despeisse (2016):

- Increasing resource efficiency through the possible redesign for AM in both production and use phases,
- Extending product lifecycle through repair, remanufacture and refurbishment.

Looking at emerging new product development processes, designers are able to reconfigure the design at a minimum of time and cost, or can make models and prototypes using the exact material of the final parts to be tested, resulting in savings in time, cost, raw materials, and overall energy consumption. One other possibility offered by AM is the minimum time required between the finalized design and initiation of production—just compare it with conventional manufacturing in which the time in needed for the designing and production of molds and fixtures, equipping the production line, and lots of routine tasks.

Further cost reductions can be found when the part is modified to fully exploit the potential of AM. There is currently a lack of education among designers and organizations which are used to following conventional design thinking. Organizations must invest in skill development and education with regard to this emerging technology so that employees will be able to recognize its full benefits and potential uses.

Cost savings in the use phase can be obtained from the biological production principle, which is that layer-by-layer approaches are inherently less wasteful than traditional subtractive methods of production. Additionally, the capability of producing integrated parts without the need for assemblies has a profound effect on the production line. Further to its quality impact, integrated products will lead to a minimizing of waste and corresponding labor costs, and facilitate a flatter and simpler supply chain thanks to fewer required components.

AM delivers many benefits to manufacturers in the production process and process design. Through incorporating 3D printed toolings (e.g. molds) that make use of forms only possible through AM, the production process has become more energy and resource efficient. An example of this can be seen at the Finnish company Salcomp, a manufacturer of electrical plugs and mobile phone power supplies. In this highly competitive industry, cost and efficiency are the major driving factors for maintaining competitive position. The company identified that a limiting factor in its injection molding process was the cooling time. In collaboration with EOS, their engineers redesigned—using DMLS technology—the vent structure of the molds used to dissipate heat more quickly. The main benefit of the redesigned mold was a reduction in the cooling time from 14 to 8 s, allowing 56,000 more units to be produced each month. It was also useful in terms of quality improvement, resulting in a reduction of rejection rates from 2.0 to 1.4% (Chen et al. 2015).

A summary of the potential sustainability effects of AM, including both benefits and challenges is in Table 4.2. The impacts are categorized based on the lifecycle

Table 4.2 The values of AM during product lifecycle

	Advantages	Challenges
Product (re)design	Enabling lightweight parts	Educated designers based on the capabilities of AM
	Product dematerialization	Supporting the skills development of prosumers
	Integrated parts	Integrating sustainability considerations using eco-design principles
	Reduced time between design and production	Learning in future applications
	Localized material recycling	
Material input processing	Improved resource efficiency of raw material processing	Recycling potential limited to certain
	Reduced toxicity of material processing	Scaling-up processes for new materials
	Upcycling of waste materials into new applications	Lack of knowledge of the environmental aspects of material processing
		Validating material properties
		Certifying materials
		Standardizing materials and processes
		Increasing percentage of recycled content in material inputs
		Limited recyclability of products at their end-of-life due to diverse materials
		Avoiding material contamination
Component and product manufacturing	Reduced energy intensity	Limited process speed
	Reduced waste	High machine costs
	Improved resource efficiency	Lower cost effectiveness and energy efficiency at higher production volumes
	Flexibility through make-to-order manufacturing	Lack of environmental knowledge and understanding of the environment
	Reduced cost of customization	Integrating AM with hybrid technologies in design and production
	Reduced material inputs and handling reduce costs	Limited automation
	Reduced inventory waste including unsold goods	Certifying manufacturing processes
	Integrated parts (no need for assembly) lead to simpler supply chains	Requirements for standards and regulations
	Raised awareness of manufacturing process and its impacts	Quality control in distributed networks
	Increased equipment utilization	Maximizing machine usage in the home by prosumers
	Automated production using hybrid technologies	Optimizing AM build process
	Higher quality finish achieved using hybrid technologies	Limited availability of digital designs
		Cost of acquiring new digital designs
		Fragmentation and uneven distribution of current AM services
		Potential for AM to contribute to a materialistic society and consumerism

(continued)

Table 4.2 (continued)

	Advantages	Challenges
		Replicating learning in future applications
Product use	Lightweight parts	Uncertain performance of products and components due to low maturity of technology
	Improved operational efficiency	
	Improved functionality and durability	Uncertain performance of products and components over extended lifespan
	Component upgrade for product life extension	
Repair and remanufacturing	Reduced waste during repair process	New business models in other sectors
	Reduced process time for repair	Implementing distributed maintenance systems
	Improved product utilization through repair and remanufacturing	Certifying repair and remanufacturing processes
	In situ repair and remanufacturing	Certifying spare parts to overcome liability issues
	In situ and spot repair	
	Component upgrade during repair process	
	Product-service business models for repair and remanufacturing	
Recycling	Increased recycling rate	Limited recyclability of plastics due to quality losses
	Use of recycled materials by-products	Non-recyclability of AM-produced multi-material goods
	Simplified assemblies with less material diversity improves opportunities for recycling	Educating consumers about recycling AM plastics
	Localized recycling systems	Incompatibility between non-standardized, non-recyclable materials
	Raised awareness of material recycling	
	Increased acceptance of recycled material content	
	Democratized material recycling	

Source Ford and Despeisse (2016)

phases starting with product design, material inputs, component and product manufacturing, product use, repair and remanufacturing, and finally the recycling phase.

As illustrated in Chap. 3, the sustainability implications of AM technologies are present throughout the product lifecycle, consisting of raw material, manufacturing, distribution in the production phase as well as the use phase. The majority of cost savings in the production phase are the result of:

- Reduced handling costs,
- Shorter supply chain,
- Fewer demands for materials,
- On-demand production,

- Lower inventory levels due to a minimum need for warehousing raw materials, work in progress and final products,
- Lower labor costs, only for pre-and post-processing due to advanced automation,
- Lower fuel consumption obtained from lightweight parts.

Moreover, AM affects the operational costs particularly the costs of flexibility, customization, new product development, capital, and marginal production.

4.2.3 Impact on Supply Chain Management

The supply chain process is defined as "the activities that produce a specific output of value to the customer" (Lambert 2014). Accordingly, as seen in Fig. 4.8, the process can be split into eight different sub-processes. Supplier relationship management is the process of maintaining and developing relationships with suppliers. Manufacturing flow management involves all the activities through which manufacturing flexibility and the flow of goods through the supply chain are manageable. Product development and commercialization is the process of collaborative product development and production with customers and suppliers. Other sub-processes involve order fulfillment, demand, customer relationship management. These are the processes of customer relationship management, the managing of product and services agreements, and all the activities from which customer order will be met. The final sub-process, namely return management includes activities in reverse logistics, for instance managing unwanted returns or managing reusable assets.

The components of the supply chain network include raw material suppliers, sub-assembly, information systems, logistics, retailers, and ultimately the customer.

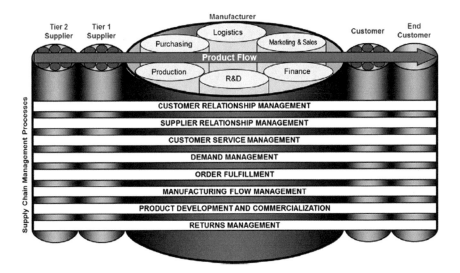

Fig. 4.8 Supply chain process framework. *Source* Lambert and Enz (2017)

Supply chain management as defined by Slack et al. (1998) is concerned with the holistic management of the supply chain as a whole. A number of supply chain approaches/paradigms have been developed in recent years. These include Lean, Agile, and Leagility. The impacts of AM on these paradigms are discussed below on the basis of the research of Tuck et al. (2007).

Lean paradigm

The lean supply chain paradigm seeks to reduce waste throughout the chain. Cox (1999) presented a summary of the eight lean supply chain characteristics as follows:

1. Perfection in delivering value to customers.
2. Produce only what is demanded from the customer through Just-In-Time and focus only on those actions that create value.
3. Concentration on the elimination of waste in all operational processes.
4. Considering all of the participants in the supply chain as stakeholders, consequently we must add value for everyone in the business.
5. Develop win-win, rather than win-lose relationships with suppliers.
6. Cooperate with suppliers to create a lean and demand-driven logistics process.
7. Reduce the number of suppliers with those given a preferred long-term relationship.
8. Create a network of suppliers to cooperate in waste reduction and operational efficiency.

Furthermore, the lean paradigm relies on some fundamental market principles in order to be more effective. Fisher (1997) stated that the lean paradigm should be suited to the products, which are long product lifecycle, low margin, low product variety, accurate forecasting of demand, and where the market winner is the one with lower cost. Using some of these eight lean supply chain characteristics helps us to better understand and distinguish the impacts of AM on the lean supply chain. AM affects two of these, which are related to the Just-In-Time (JIT) concept, and reducing waste in the whole supply network.

As we will see in Sect. 5.2, the implementation of AM will affect JIT manufacturers through:

- Dematerializing Supply Chain
- Just-In-Time manufacturing
- Reducing setup and changeover time and cost, and number of assemblies
- Reducing waste.

Agile and Leagile paradigm

The agile paradigm focuses on reducing lead time rather than waste. Naylor et al. (1999) define agility as follows: "Agility means using market knowledge and a virtual corporation to exploit profitable opportunities in a volatile market place."

In fact, agility seeks flexible production methods, allowing for fast reconfigurations in order to meet consumer demands. Consequently, products with a short lifecycle, such as fashionable goods, are well suited to the agile paradigm, in

Fig. 4.9 Modified product and supply chain matrix. Adapted from Tuck et al. (2007)

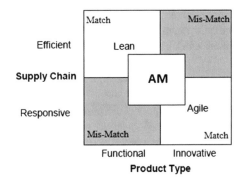

contrast to the lean paradigm's objective of commodity production. The concept of leagility involves the concepts of both lean and agile supply chains in order to obtain greater value from the supply chain. The advent of AM promotes a lean production in a responsive manner, which covers the focus of both the lean and agile paradigms. In other words, considering Fisher's 2 * 2 Matrix, the current mismatch area within an efficient supply chain and innovative product can be removed thanks to AM (Tuck et al. 2007).

The modified matrix considering the advent of AM is shown in Fig. 4.9. Broadly speaking, AM offers the following impacts on the agile supply chain: truly "leagile" supply chain; production with fast re-configurability and fast response; reduction in inventory levels; reduced waste; increased value; reduced logistics cost; reduction in part count; and increased flexibility.

Supply chain configuration and manufacturing logics

Conventional manufacturing can be considered as having manufacturer-centric approaches, whereas value creation activities are carried out on the manufacturing site and are less related to the consumer. In contrast, the implementation of AM technologies for consumer goods offers a new logic, namely a consumer-centric approach. A fundamental driver for the consumer-centric approach is the direct co-creation with users, in which an innovative design can be created whether by consumers or from crowd sources (3D design communities). In this model, the manufacturer can also establish its own platform for users in order for them to customize their own parts and ship them on-demand. This model has the additional benefit that the manufacturer can also use the models created by users as the inspiration for novel product designs and as a source of market data gathering. These all improve customer relationships and lead to greater customer satisfaction.

On the other hand, the supply chain configuration approaches (centralized and distributed) can be also adapted to consumer goods. Centralized manufacturing refers to 3D printing of objects through different manufacturers' facilities around the market. However, decentralized manufacturing will break this configuration into localized 3D printing services. Nevertheless, producing some special parts at lower production volumes still seems viable for centralized manufacturing. Decentralized supply chain configuration in 3D printing of consumer goods can be a range of

Fig. 4.10 A framework for
introducing AM based on
supply chain configurations
and business models. *Source*
Bogers et al. (2016)

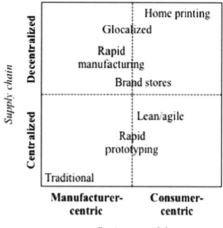

brand stores around the world, whereas they are located closer to the consumers to print parts based on the real demand. This would also lead to the use of local input in terms of material and workforce. This decentralized supply chain configuration can also occur at the point of use in the home or office, whereby the consumer can also produce by using home-use 3D printing services, promoting prosumption rather than consumption. Therefore, manufacturers may also supply these facilities or provide a platform for home printing. This will also raise the possibility for the manufacturer to outsource the production of specialized products to the consumer. Considering both supply chain configurations (centralized and decentralized) and both manufacturing logics (manufacturer-centric and consumer-centric) for consumer goods, Fig. 4.10 illustrates the positions of different approaches of adopting AM technologies (Bogers et al. 2016).

Various elements participate in the widespread application of AM technology for consumer goods, including prosumers (see Sect. 4.3.1), low-cost AM machines (see Sect. 1.5), and 3DP online services (see Sect. 4.1.3).

4.3   The Value for Customers: From Mass Customization to the New Entrepreneurial Role of Prosumers

From the customer's point of view, AM provides numerous opportunities. It offers fully customized products, and customer needs and requirements can be met by creating products that fit in color, form and function because AM offers designers an almost unlimited freedom of design. The customization of the part and faster delivery to the customers led to higher customer satisfaction. Furthermore, AM provides access to new markets and gives firms competitive advantages, which result in the acquisition of a broader range of customers. AM is a powerful tool that offers new forms of competitiveness and can better satisfy the requirements of different users, leading to a real mass customization of consumer goods and a quick response to business opportunities.

In the traditional product development process, companies often employ a sequential procedure with various groups or individuals at different stages. However, the successful product design needs proper communication throughout the creation process among all beneficiaries. In fact, the so-called phenomena of "over the wall" designing is no longer a viable means of fulfilling current and future consumer needs.

In contrast, AM provides the possibility to systematically involve the customer during all the stages of product development, manufacturing and distribution. Currently, with the help of technological advancements in providing available and affordable resources, AM has been employed in various marketplaces beside the industrial sector. The enablers are wider range of usable raw material, cheaper AM machines, and online 3DP service systems. These provide the customer with different routes by which to reach the final required product. Customers can decide whether to produce themselves in their home or office, or they can order production from available 3D printing platforms. They can also participate in AM communities and share designs and collaboratively develop a product.

Thus, using AM for consumer goods has also resulted in a shift in the current business model. The two important impacts of AM in customization together with a flatter supply chain bring huge opportunities for consumers and consumer goods. In fact, the capability of AM in freeforming and social manufacturing has shifted the adding of value from manufacturers to consumers. No longer do manufacturers act as the only source of innovation and productivity—consumers are now involved in boosting productivity too. This fact, as reported in Sect. 4.2.3, has led to different supply chain configurations driven by a consumer-centric approach.

4.3.1 Prosumers

The advent of the internet brought not only numerous benefits for consumers but also it now enables users to produce content. This has specifically been visible since the emergence of Web 2.0 technologies. In these technologies (e.g. Twitter, Facebook, Instagram), the value is created mostly by users, and the users in this environment becomes producers. Therefore, this participation unites consumers and producers, and since users can both produce and consume, they became prosumers. Prosumption takes place in a situation where consumers produce their own requirement, or consumers participate in any value chain during the production process.

AM technologies have also brought opportunities for prosumption, in which consumers become producers through the use of affordable AM equipment, or they can participate in the design and customization of objects, or be considered as a source of innovation through crowd production and co-creation. In fact, prosumers

Fig. 4.11 Consumer 3D printing categories. *Source* Kietzmann et al. (2015)

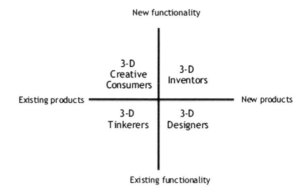

of 3DP can be involved in all supply chain tasks including design, manufacturing and distribution. The low-cost 3D printing used in the home (so no need for distribution) enables individuals to become producers, increasing the influence of consumer involvement in production and distribution. In addition, the option of co-designing using the crowdsourcing of design has had a major influence on consumer involvement in the design process.

Prosumer activities can be separated into two different dimensions: firstly, the degree to which they print either an existing product or new one, and secondly, the degree to which their 3D printed parts sustain or improve the functionality of the original part. Taking into account these two dimensions, we have four different types of prosumers working on 3D printing areas (see Fig. 4.11), namely 3D thinkers, 3D designers, 3D creative consumers, and 3D inventors (Kietzmann et al. 2015).

3D thinkers are those who work on the existing products and can maintain product functionality. An example of this work could be replacement parts for a home appliance, whereas 3D creative prosumers seeks to adopt, modify and transform existing original products to improve associated experiences. Figure 4.12

Fig. 4.12 Duplo to Brio converter brick

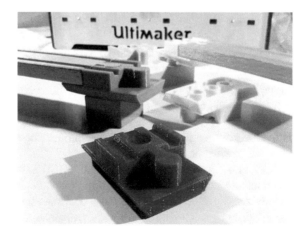

Fig. 4.13 A 3D printed hand prosthetic. *Image Source* http://www.pixabay.com/

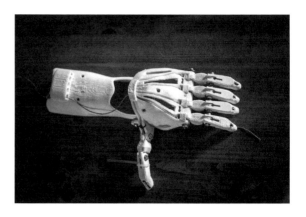

illustrates an example of this modification to the original part. This 3D printed object works as a connector between Duplo bricks from Lego and Brio wooden train tracks. With the original it seemed to be rather difficult for three-year-old children to put the track on, so a user modified the object using an open-source 3D printing service to obtain a slightly simpler design.

3D designers refer to those people who create a new product without any significant improvement in functionality. For instance, hobbyists may only fabricate a new design of coat hooks or doorstops for fun, while professionals may work on new designs of jewelry and fashion goods, and 3D inventors seek both new designs as well as new functionalities and experiences—for example, the Handie 3D printed artificial hand (e.g., Fig. 4.13) prosthetic that offers a large degree of functionality to amputees. With the help of 3D printing and a customized mechanism for finger flexing, the team made a much cheaper prosthetic hand. Prosthetic limbs like this normally cost tens of thousands of dollars, however, Handie can be fabricated at an affordable price of no more than $400. It senses brain impulses in order to control movement. Using a smartphone, Handie calculates the electrical impulses on the skin's surface to grab objects of different shapes.

4.3.2 User Entrepreneurship Opportunities

Entrepreneurship is principally dealing with either generating opportunities or recognizing and exploiting these opportunities. User entrepreneurship is defined as the commercialization of a new product or service by an individual or group of individuals who are also users of that product or service. Thus, users have the potential to become entrepreneurs.

Section 4.1.3 discussed the potential of 3D printing for providing emerging marketplaces with opportunities for entrepreneurship. These marketplaces can be categorized into four types. First, those providing sophisticated 3D designs. To run the 3D printing from elementary users to industrial ones, specific design data are needed to meet the technology's requirements. Second, 3D printing users and designers need intermediaries to be linked together in order to collaborate in the creation of designs. Consumers may request a special design or part, and designers provide them with the information. Third, prosumers need to obtain their means of production such as a 3D printer, material and accessories for printing—online or local stores should supply this hardware. In addition, since such hardware may be unaffordable or not justifiable for consumers, some marketplaces offer different services providing consumers with an easier way. They may provide 3D printing services, in which consumers can upload their design and the 3D printed object will be shipped to them, or consumers may rent 3D printing for a period at a Fablab. Fourth, consumers can commission a 3D printing service to oversee the entire product development process from the initial design to the final physical product.

Further to the above-mentioned opportunities for entrepreneurship, the industry also needs consultants during the entire process. Since 3D printing enables production from beginners to professionals, it is conducive to fostering user entrepreneurships. Of course, it will improve further with advancements in technology, and those barriers are discussed in Sect. 4.4.

The study by Holzmann et al. (2017) proposed a framework through which to realize the opportunities for user entrepreneurship. The framework (Fig. 4.14)

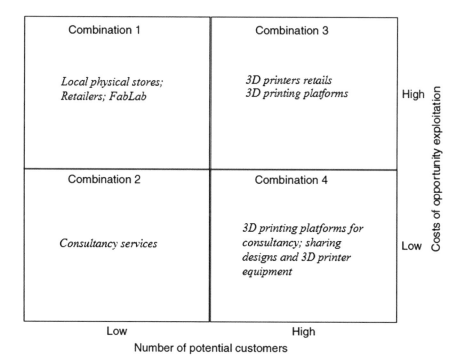

Fig. 4.14 A framework for user entrepreneurship opportunities with 3D printing. Adapted from Holzmann et al. (2017)

analyzes business opportunities based upon their cost of exploitation and quantity of potential customers. As can be seen in the figure, there are four combinations of "number of potential customers" and "cost of opportunity exploitation".

Low customer numbers and higher cost of opportunity exploitation

Obviously, the least attractive businesses are those that need higher initial investment, while having few potential customers in future. Therefore, in this situation the user entrepreneurships has negative feedback and implementing 3D printing would be less advantageous. These 3D printing services can be those that supply things in physical stores and owning themselves all the elements of the manufacturing. Local 3D printing services, local retailing of accessories, and operating a local fablab belong to this category. In these cases, they will have a limited number of potential customers, while having higher fixed costs. For instance, they have to purchase a varied range of equipment and provide trained employees and workstations in order to fulfill customer requirements.

Low customer numbers and lower cost of opportunity exploitation

This combination cannot be considered as a very attractive opportunity. The lower cost of opportunity exploitation compared to the expected gains results in a lower risk option for user entrepreneurs. An example of this situation in the 3D printing market could be consultancy services provided to consumers and prosumers. The service can provide assistance in 3D design, selecting technology and materials, operating, and provide training courses to meet the needs of users. This group of user entrepreneurs will have higher marginal returns since they do not have to invest a lot of capital.

High customer numbers and higher cost of opportunity exploitation

This combination can be considered as an attractive opportunity for user entrepreneurs, however very careful consideration of the advantages and disadvantages is needed. Two businesses related to 3D printing belong to this combination, namely online 3D printing retailers and online 3D printing platforms. The former usually retails 3D printers, their consumable components, accessories and material. Considering that 3D printers are currently becoming more affordable for users, this business market is appealing. Moreover, the fixed costs have been reduced since they are offering online retailing. However, the competition seems to be high as even leading retailers like Amazon or eBay are selling 3D printers. Therefore, user entrepreneurs must cut their costs to compete and this can be achieved by, for instance, purchasing a large number of 3D printers and other required items.

The second business opportunity is 3D printing platforms (for further discussion see Sect. 4.1.3). Currently, several online 3D printing platforms share and sell designs. However, a design-hosting platform can be attractive to users if it provides a steadily increasing number of new designs. Similarly to the 3D printer retailing market, this market is also very competitive as there are several well-stablished platforms which already have thousands of users. Moreover, user entrepreneurs have to spend a significant amount of money to set up a platform and promote it.

High customer numbers and lower cost of opportunity exploitation

This combination is the most promising opportunity for user entrepreneurs who can benefit from 3D printing capabilities in this situation. This can be obtained through the use of online services, in which it is easier to find a broader range of customers while needing a lower cost of opportunity exploitation and less time for launching the service on the market. Thus, it is expected that using this opportunity will lead to a shorter return on investment (ROI) period.

As for 3D printing markets, there are some business opportunities for user entrepreneurs in this area. First, users may use online services for consulting on customers' needs. Second, experts in computer aided design software can sell their designs whether through their own websites or by sharing on 3D printing platforms for a fee. The platform lets designers share their designs so that they are accessible to a greater number of potential customers. Designers then set the price for downloading a design and the platform will charge a commission fee on every download. Third, user entrepreneurs who own 3D printers and have design skills can offer online 3D printing services for making objects or rapid prototyping services.

4.4   Strategic Challenges and Barriers Ahead of AM Technology Diffusion

The benefits and advantages of AM technologies are often discussed such as freedom of design, integrated parts, toolless production, on-demand production, and many other opportunities. Considering those in the round, one may expect the diffusion of AM technologies to be similar to that seen with the internet today, but this is not the case. In fact, according to Wohlers (2014), AM still contributes less than 2% of the whole manufacturing market. Hype cycle is a tool to analyze the status of a specific technology. The hype curve is used to depict the maturity, adoption and social implication of a specific technology. Even though AM technologies passed their plateau of productivity in 2015 (Gartner Inc. 2015), according to Huang et al. (2015) considering the wide public awareness and more distributed industrial interests. A second hype curve for the technology should be taken into account. Therefore, in this case, AM is still in the developing phase of being a

Fig. 4.15 AM Technology lifecycle. *Source* Chu and Su (2014)

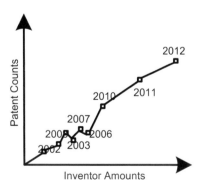

widespread method in the mainstream manufacturing area. In fact, when the general public acquires a good understanding of the technology it may experience a downward trend, likely in the near future.

Moreover, studies demonstrated that the technology is still in the growth stage. Thus, it is crucial that technologists and expert communities begin a substantial dialogue regarding these challenges to enable the technology to become much more available to mainstream markets that currently have little to no need for AM. It is becoming an increasingly mature and professional method due to consideration of the potential for technological advancements. Technology life cycle analysis is a useful method to understand the status of the technological trends. This method considers the patents and inventors amount per year. As for additive technologies, Chu and Su in 2014 conducted a research on the status of AM technologies. They demonstrated that the technical part of AM is dramatically growing since 2007. Figure 4.15 is an adapted diagram form their study, illustrating the trend of technological advancements based on the patent counts and inventor amounts.

There are some gaps in the development path, which AM suppliers and communities should consider. This potential for advances and widespread applications is categorized by machinery, design tools, process and materials (for AM supplier), while two others are mostly related to AM communities, namely education and standardizations (see Fig. 4.16).

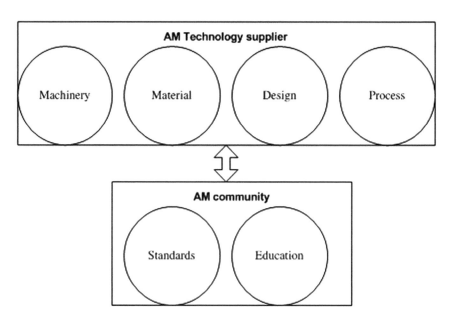

Fig. 4.16 Dimensions of AM adoption barriers

4.4.1   Challenges Ahead of AM Technology Suppliers

Considering AM suppliers, there are several barriers to a higher level of AM technology diffusion, similar to those for other technologies at different stages of maturity. Identifying and addressing these barriers would help users and inventors to improve the performance of technology and increase the adoption of AM.

Machinery

A major challenge facing AM machines is having the ability to produce large-sized parts. The ideal manufacturing method should be able to produce an unlimited product size from the smallest possible to the largest and efforts can be seen in this direction in the construction of real houses using 3D printing, however, it remains a challenge, for instance in tooling processes that usually require large size molds and manufacturing tools. Therefore, in order to develop the range of applications, AM machine manufacturers and research institutions should focuses on creating a larger viable build chamber.

As with other technologies, the ease of use is an important factor affecting uptake of the technology (mostly for home-use 3D printing). Although 3DP suppliers attempted to make printers more reliable and straightforward to use, the ease of use of printers and software, particularly in terms of the graphical user interface (GUI), became a limiting factor in terms of reaching a broader market. Generally, 3DP manufacturers need to take responsibility for end-to-end operability by increasing the integration of all the elements of the system.

Moreover, in the initial stage of development, most materials were exposed to shrinkage or curling, and the production throughput was therefore limited by the speed of the drying process. It also could not properly support complex cantilevered designs as the layers are created, thus new software is required to work around the mechanical limitations of in-progress production. In terms of operational costs, it was initially too slow and resulted in more expensive parts at medium and high volume production. Due to these limitations, the initial application concentrated on rapid prototyping and tooling, revealing that for wider applications and the use of AM for end-use products, significant developments are still needed. In addition, the price of printing machines was another constraint on widespread usage that limited the application sectors to medium and larger companies. In industrial sectors and usually for small enterprises, the high initial investment is the main barrier to the adoption process. In addition, for home-use (desktop) 3D printing, a much more concerted effort should be made to supply more affordable machines with a higher level of quality.

Materials

Usually, the development of an alloy in metallurgy takes about ten years, due to the need to fully understand the various properties such as fatigue strength. Therefore, this time is also required for new AM materials, which depend on various features such as greater mechanical properties or higher corrosion resistance. This is the case in the aerospace industry where materials have to resist staggering pressures and temperatures of over 1000 °C. In fact, one of the key drivers of the spread of this technology is the development of a much larger range of materials. For instance, using composite fillers can be a relatively simple way to extend the range of materials available.

Currently, metal is considered as the most exciting material for AM technologies, from which steel, titanium and cobalt chromium can be noted as the most widely used (titanium is widely used in both the aerospace and medical component industries). Moreover, Aluminum has also been introduced to this range of materials, along with platinum, brass, gold, silver, and copper. Even for the currently available materials, further advances should be taken into account such as developing desired metals or other materials in inks to satisfy resolution requirements down to sub-micrometers. Moreover, many of these materials may need additives to improve the fusion process (for sintering). Ceramics also still need to be developed as in spite of many publicized glass printers, the outcomes do not satisfy the needs of some application sectors such as optical goods.

Design

Initially, the customers of 3DP were those manufacturers seeking rapid prototyping. 3DP manufacturers designed their systems to use existing application software, particularly CAD (computer-aided design software). Therefore, in order to access these workflows, AM equipment suppliers were required to operate with such CAD programs. The STL (STereo-Lithography) file format, created in 1987 by 3D Systems, became the standard for digitally defining the surface of a 3D object using a series of triangular facets. STL provided a mechanism for communication between CAD and other 3D graphic applications and drivers that rendered such designs into physical objects. It was rapidly adopted as the common format for 3DP and remains in use thirty years later.

The inclusive capability of AM in offering freeform fabrication allows designers to explore new applications of this technology and the manufacture of end-use parts. Therefore, to gain full advantage from this feature new design tools should be developed to aid the use of these new functions and capabilities. Different novel design tools are needed, such as for conceptual designs that should represent the AM design space, process, and properties; tools for integrated product and process design, and methods for assessing lifecycle costs. Current CAD software should also be adapted to encompass the limitations of 3D designs of complex geometries, and to move forward to a more user-friendly environment, for beginners as well as for professional designers.

As for home-use 3DP, the availability of content is another barrier to its adoption. For industrial buyers, the designer usually generates the digital designs. However, when using the technology in a smaller market place (namely in the home or office) available digital design files are needed by beginners. 3DP manufacturers, together with other suppliers, provide catalogs which contain digital designs of standardized physical goods (such as nuts and bolts), however, the currently available services are not helping to grow the application of the technology.

Additionally, according to Ilyas (2013), the combination of an advanced manufacturing technology, called 3D Machine Vision (3D-MV), with existing AM technologies, provides additional profits and benefits with which to accelerate the product development process. The scanning process is deployed in reverse engineering, the objectives of which include the realization of a given product design and its details. In other words, it generates the concept model from the existing physical model.

The reverse engineering process can be divided into three main phases. In the first phase—the scanning process—3D-MV is utilized to capture and record the data of the given product's surfaces in a numerical format in order to create a matrix of 3D coordinate points. The second phase—the modeling process—corrects and modifies the data. Scanning data always requires some modification (namely reducing the number of points, fixing scanning results, or the merging of different scanning results) in order to provide a replica. Then, the modified data are imported to a reconstruction software package for conversion to CAD. The third phase—model verification—is to verify the whole processes through quality inspection of the product and any redesign requirement.

The combination of these two advanced manufacturing technologies (3D-MV and AM) has a positive effect on the process and product. It provides proper and rapid technical data relating to the object, resulting in an acceleration of the design process through improvements in design documentation. It also overcomes a limitation of CAD in relation to product modification and improves its manufacturing process. More importantly, it is a simpler way of redesigning a part in pursuance of improving the performance and quality. It is obviously beneficial for the manufacturing of a part when there is no available drawing and manufacturing documentation available. On the other hand, further advances and developments such as the integration of these two technologies in terms of hardware and software are needed to make it justifiable in terms of cost and to make it technically usable.

Process

Most manufacturers are cautious about implementing AM as an alternative manufacturing method due to the repeatability and durability of the manufactured parts. For instance, in conventional manufacturing, producers receive dynamic feedback during the fabrication of parts. In contrast, AM manufacturers do not use a closed loop system for getting immediate feedback and consequently some manufacturers are skeptical of the structural integrity of the finished product in comparison with conventional manufacturing (Sealy 2012). The problem of accessing the build chamber could be considered as a challenge to process control and quality inspections. It may require faster tools located in the build chamber in order to sense temperature, geometric dimensions, surface quality, and cooling rate. Moreover, post-processing increases the amount of manual operations, which may lead to operator errors, increasing the total process time and costs. Thus, further advancements may be accrued in automated post-processing, boosting its efficiency.

To sum up the challenges facing the technology, the results of a survey will be discussed. Several barriers to the adoption of AM were identified through a survey of 700 professionals and AM users, conducted by Stratasys (a leading 3DP manufacturer) in 2015. As seen in Fig. 4.17, the results demonstrate that two major barriers both now and in the future are perceived to be the cost of equipment (mostly industrial-grade equipment) and the limitation of the material range, with two other factors being less impactful, namely the cost of production and the finishing part of process.

Notably, apart from these four main barriers, other significant factors that currently exist should be taken into account as follows:

- Lack of in-house additive manufacturing resources
- Lack of expertise and/or training among workforce/employees
- Limited repeatability (accuracy from build to build)
- Lack of formal standards
- Lack of proven documentation of additive manufacturing's capabilities
- Software development and capabilities
- Longer production timelines
- Limited recyclability
- Risk of litigation/legal implication
- Data storage requirements
- Accuracy and surface finish
- Full color capabilities.

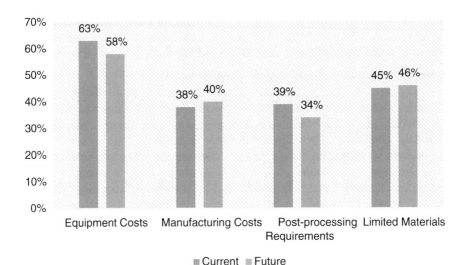

Fig. 4.17 The main barriers to the adoption of AM technologies.

4.4.2 Challenges Facing the AM Community

Although technological barriers exist, the majority of barriers tends to be non-technical and instead are cultural and human-centric issues. For instance, the lack of standards and an educational plan does not fit the development status of the technology.

Standards

Certifying and standardizing the components of production, including equipment, materials, processes and logistics, is an essential factor for consistency in manufacturing. Currently, the lack of confidence in manufactured parts in terms of their mechanical properties inhibits widespread take up of the technology. Most industry experts agree that a lack of standards is a key barrier to its adoption. In fact, not only are the existing standards not sufficient for AM, but they also do not respond well to the specific characteristics of AM. This lack of standards can be identified as a challenge in the following specific areas (Monzón et al. 2015); materials, process and equipment, qualification and certification, and modeling and simulation.

Nevertheless, two leading standards associations, the ISO (International Organization for Standardization) and ASTM International (American Society for Testing and Materials) have assigned committees and prepared some standards. The F42 committee of the ASTM was established in 2009 to work on additive manufacturing. This committee involves more than 200 experts from sixteen countries, mainly from the USA and Europe. The ASTM committee is working to develop standards for the following objectives:

- Allowing a comparison of the performance of different additive AM technologies and materials
- Specifying fabrication requirements to give purchasers and suppliers a common set of parameters
- Guiding early adopters to use AM technologies with a uniform procedure.

The F42 committee comprises several subcommittees such as F42.01 Test Methods; F42.04 Design; F42.05 Materials and Processes; F42.90 Executive; F42.91 Terminology; and F42.94 Strategic Planning, F42.95 US TAG to ISO TC261. A list of key standards is presented in Table 4.3 (ASTM, June 2014). Moreover, the ISO committee TC 261 on AM technologies was also established in 2011. In addition, these two standards associations have decided to collaborate in order to develop the standards for the following areas:

- Design guidelines
- Standard test artifact
- Requirements for purchased AM parts
- Harmonization of existing ISO/ASTM terminology standards
- Standard specification for extrusion-based plastic materials.

Table 4.3 Key ASTM standards for AM (ASTM, Jun 2014)

Committee	Standards
F42.01. Test methods	F2971 Standard Practice for Reporting Data for Test Specimens Prepared by Additive Manufacturing
F42.05. Materials and processes	F2924 Standard Specification for Titanium-6 Aluminum-4 Vanadium with Powder Bed Fusion
	F3049 Standard Guide for Characterizing Properties of Metal Powders Used for AM Processes
	F3055 Standard Specification for Nickel Alloy (UNS N07718) with Powder Bed Fusion
	F3056 Standard Specification for Nickel Alloy (UNS N06625) with Powder Bed Fusion
	F3091/F3091M Standard Specification for Plastic Materials With Powder Bed Fusion
F42.91. Terminology	F2792 Standard Terminology for AM Technologies
	F2921 Standard Terminology for AM; Coordinate Systems and Nomenclature
F42.04. Design	F2915 Standard Specification AM File Format (AMF)

Education

In the case of AM, there is still a lack of education and cultural differences exist. Based on information reported from The Manufacturing Institute (USA) in 2014 (cited from Huang et al. 2015), 83% of US manufacturers outlined the shortage of qualified employees in additive manufacturing. In a similar way to other emerging technologies, the AM industry needs to develop its own education and training programs. It requires training for current employees and education in schools and at university level for the future workforce. These education programs should not only involve different relevant practitioners such as machine operators, product designers, R&D engineers, and business administrators but should also cover public communities such as political leaders and policymakers, individual entrepreneurs and economic development institutes.

Exhibit 6—Current Barriers to the Adoption of AM Technologies—Is AM Technology Accessible or Is It Still Too Expensive?
There are some gaps in the development path, which AM suppliers and communities should consider if further advancements are to be made resulting in a broader adoption of the technology. Among these barriers, the cost of equipment and the limited range of available raw materials are considered as the most significant factors.

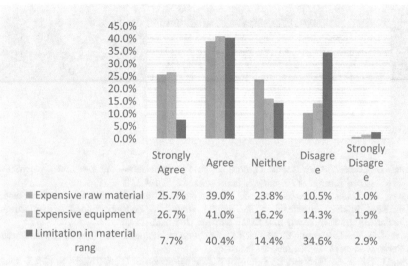

	Strongly Agree	Agree	Neither	Disagree	Strongly Disagree
■ Expensive raw material	25.7%	39.0%	23.8%	10.5%	1.0%
■ Expensive equipment	26.7%	41.0%	16.2%	14.3%	1.9%
■ Limitation in material rang	7.7%	40.4%	14.4%	34.6%	2.9%

The results of our survey show that more than 67% of the companies believe that the cost of equipment in hindering widespread adoption of the technology. Approximately 65% of respondents agree that raw materials for AM are more expensive than those in conventional manufacturing. In addition, more than 48% of respondents agreed that the range of raw materials is still limited. Therefore, many potential adopters avoid investing in this technology as a direct result of these factors. However, in prototyping or mock-up development, it is possible to use cheaper raw materials than in the actual object, as well as using low-cost 3D printers.

Seven Key Facts

- Driving company innovation and individual creativity is one of the major implications of AM technology.
- AM can be considered as a disruptive technology when it is adopted by manufacturing companies that have relied on conventional manufacturing.
- Online 3D printing platforms supply designs, host designs, provide co-designing, crowdsource designs, 3D print objects, sell 3D printers and/ or crowdsource 3D printing.
- AM is a cost effective and technically feasible method for customization (high variety products with low production volumes) due to the elimination of molds and any extra tools.
- AM leads to lean manufacturing in a responsive manner.
- Manufacturers no longer act as the sole source of innovation and productivity, while consumers are becoming involved in boosting productivity.

- Current technological barriers to adoption are categorized as machinery, design tools, process and materials for AM suppliers and education and standardization for AM communities.

References

Alford, D., Sackett, P., & Nelder, G. (2000). Mass customization—An automotive perspective. *International Journal of Production Economics, 65*(1), 99–110.

Bogers, M., Hadar, R., & Bilberg, A. (2016). Additive manufacturing for consumer-centric business models: Implications for supply chains in consumer goods manufacturing. *Technological Forecasting and Social Change, 102,* 225–239.

Bower, J. L., & Christensen, C. M. (1995). Disrupting technologies: Catching the wave. *Harvard Business Review,* 43–53. January-February.

Chen, D., Heyer, S., Ibbotson, S., Salonitis, K., Steingrímsson, J. G., & Thiede, S. (2015). Direct digital manufacturing: definition, evolution, and sustainability implications. *Journal of Cleaner Production, 107,* 615–625.

Chu, Y. T., & Su, H. N. (2014, July). Understanding patent portfolio and development strategy of 3D printing technology. In *Management of Engineering & Technology (PICMET), 2014 Portland International Conference on* (pp. 1407–1415). IEEE.

Cox, A. (1999). Power, value and supply chain management. *Supply chain management: An international journal, 4*(4), 167–175.

Fisher, M. (1997). What is the right supply chain for your product? *Harvard Business Review, 75*(2), 105–116.

Ford, S., & Despeisse, M. (2016). Additive manufacturing and sustainability: An exploratory study of the advantages and challenges. *Journal of Cleaner Production, 137,* 1573–1587.

Fox, S. (2003). Recognising materials power. *Manufacturing Engineer, 82*(2), 36–39.

Gartner Inc. (2015, August). *Hype Cycle for emerging technologies.* http://www.gartner.com/newsroom/id/3114217.

Holzmann, P., Breitenecker, R. J., Soomro, A. A., et al. (2017). User entrepreneur business models in 3D printing. *Journal of Manufacturing Technology Management, 28*(1), 75–94.

Huang, Y., Leu, M. C., Mazumder, J., & Donmez, A. (2015). Additive manufacturing: current state, future potential, gaps and needs, and recommendations. *Journal of Manufacturing Science and Engineering, 137*(1), 014001.

Ilyas, I. P. (2013). 3D machine vision and additive manufacturing: Concurrent product and process development. In *IOP Conference Series: Materials Science and Engineering* (Vol. 46, No. 1, p. 012029). IOP Publishing.

Kietzmann, J., Pitt, L., & Berthon, P. (2015). Disruptions, decisions, and destinations: Enter the age of 3-D printing and additive manufacturing. *Business Horizons, 58*(2), 209–215.

Khorram Niaki, M., & Nonino, F. (2017a). Additive manufacturing management: a review and future research agenda. *International Journal of Production Research, 55*(5), 1419–1439.

Khorram Niaki, M., & Nonino, F. (2017b). Impact of additive manufacturing on business competitiveness: a multiple case study. *Journal of Manufacturing Technology Management, 28*(1), 56–74.

Lambert, D. M. (2014). *Supply chain management: Processes, partnerships, performance* (4th ed.). Sarasota, FL: Supply Chain Management Institute.

Lambert, D. M., & Enz, M. G. (2017). Issues in Supply Chain Management: Progress and potential. *Industrial Marketing Management, 62,* 1–16.

Love, D. (2014). *An Illinois high school is 3D printing a new hand for a girl born without fingers —And it will only cost $5.* Retrieved from www.businessinsider.com/kylie-wicker-boylan-high-school-3dprinted-hand-2014-4.

Marquis, D. G. (1969). The anatomy of successful innovations. *Innovation, 1*(7), 28–37.

Monzón, M. D., Ortega, Z., Martínez, A., & Ortega, F. (2015). Standardization in additive manufacturing: Activities carried out by international organizations and projects. *The International Journal of Advanced Manufacturing Technology, 76*(5–8), 1111–1121.

Naylor, J. B., Naim, M. M., & Berry, D. (1999). Leagility: Integrating the lean and agile manufacturing paradigms in the total supply chain. *International Journal of Production Economics, 62*(1), 107–118.

Rayna, T., Striukova, L., & Darlington, J. (2015). Co-creation and user innovation: The role of online 3D printing platforms. *Journal of Engineering and Technology Management, 37,* 90–102.

Sealy, W. (2012). Additive manufacturing as a disruptive technology: How to avoid the pitfall. *American Journal of Engineering and Technology Research, 11*(10).

Slack, N., Chambers, S., Harland, C., Harrison, A., & Johnson, R. (1998). *Operations management* (2nd ed.). Pitman Publishing.

Slack, N., & Lewis, M. (2003). *Operations strategy.* Upper Saddle River: Prentice Hall.

Tuck, C., Hague, R., & Burns, N. (2007). Rapid manufacturing: Impact on supply chain methodologies and practice. *International Journal of Services and Operations Management, 3*(1), 1–22.

Tuck, C., Hague, R., Ruffo, M., Ransley, M., & Adams, P. (2008). Rapid manufacturing facilitated customization. *International Journal of Computer Integrated Manufacturing, 21*(3), 245–258.

Wohlers, T. (2014). *Wohlers report 2014: 3D printing and additive manufacturing state of the industry.* Fort Collins, CO, USA: Wohlers Associates Inc.

Steenhuis, H. J., & Pretorius, L. (2017). The additive manufacturing innovation: A range of implications. *Journal of Manufacturing Technology Management, 28*(1), 122–143.

Chapter 5
The Value for Operations

Three fundamental processes in manufacturing organizations are new product development, manufacturing (processing of materials) and logistics (material handling). Nevertheless, the effects of the introduction of AM technologies exist outside the company boundaries and influence the whole of the supply chain. This chapter begins by discussing the value for the new product development process, and goes on to explain the new design methodology for AM technologies and the exclusive benefits that AM brings to the design process. It continues by discussing the value for manufacturing and considering in detail the impacts on the production process, product quality, and operational costs. The chapter ends by looking at the value for logistics and supply chain management. This section includes the impacts of AM on inventory turnover, planning for spare part supply chain configurations, and designing supply chains for 3DP platforms.

5.1 The Value for New Product Development

Almost all the activities dealing with design are performed in a CAD interface. These activities may involve the designing or redesigning of a product to incorporate all of the advantages offered by AM technology (reduced weight, complexity, integrated functionalities, and so on). Moreover, 3D data can be obtained from the physical model through reverse engineering (scanning of the physical object).

5.1.1 Impact of AM on the Product Development Process

The following paragraphs compare the impacts of AM technology on the conventional product development process. The impact of AM and its role in the

© Springer International Publishing AG 2018
M. Khorram Niaki and F. Nonino, *The Management of Additive Manufacturing*,
Springer Series in Advanced Manufacturing,
https://doi.org/10.1007/978-3-319-56309-1_5

product development process can be clearly identified through looking at the steps involved in the conventional design process.

The first step in conventional design, which is the generation of detailed drawings, can be easily eliminated. Because AM uses 3D CAD data sent directly to the manufacturing machine, it does not need any additional details of the drawing and dimensions, etc. However, in the AM process, CAD data need only to be transformed to a machine-readable format (e.g. STL). The second step in the conventional product development process is the creation of a prototype. Conventionally, prototyping departments involved multifunctional teams such as machinists, artisans, experts on mechanics, design, electronics and so on. Due to AM's capabilities, prototyping units need only trained personnel. In addition, prototypes were traditionally expensive due to the need for tools and molds. Rapid prototyping (RP) is the collective name for a set of AM technologies, used to manufacture models (prototypes).

A traditional prototype (e.g. hand-made models) had limitations with respect to functionality evaluation. Although, these models could represent the aesthetics and shape of a final product, adding functional characteristics to the model was difficult or impossible. AM provides for powerful communication between designers and customers during the product development process, due to the capability of the technology in combining aesthetics and functionality into the prototypes. The role of functional prototypes in enhancing customer interactions is illustrated in Fig. 5.1. The figure depicts a design philosophy where functional prototypes are achievable and are the central elements of a highly iterative process. Customer input can be recognized at each stage of the product development process through experimentation on the unfinished products. However, customers' input in conventional design process are only the initial requirement statement.

Fig. 5.1 The role of functional prototypes in customer interactions. *Source* De Beer et al. (2009)

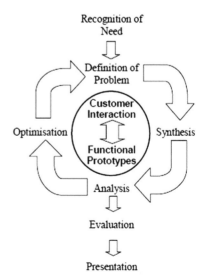

Thus, the introduction of AM radically affects prototyping scope though a simple and affordable fabrication technique. With the rapid growth and development in the AM industry in terms of materials and machine characteristics, it is reasonable to assume that in the near future AM will be the first and most sustainable choice for prototyping. In this instance, AM can be also a disruptive technology to the current available prototyping technology.

The third step in the conventional product development process is numerical control programming and tooling design. This step is normally a time-consuming and lengthy process. Using AM, manufacturers do not need to create a Computer Numerical Control (CNC) program, either for tool production or for manufacturing purposes. There is also no need to select tools, designing and producing fixtures, or test tools, etc. In addition, the capital investment, considering the efforts being made to reduce the price of AM machines, is definitely less than that in conventional manufacturing such as computer-aided manufacturing (CAM) and CNC machines (Mansour and Hague 2003).

A majority of the companies benefited from the implementation of AM in the process of new product development (khorram Niaki and Nonino 2017a). These include model making aimed at fitting and functional tests, as well as making mock-ups which are similar to the final products in terms of material and color. It obtains appropriate feedback, resulting in improvement in designs and boosts in its effectiveness. It also offers a full customer relationship before the final product stage, ensuring customer satisfaction through meeting their requirements. AM also provides for cost-effective design modification, thereby increasing the flexibility and changeability of the company to meet customer needs. In addition, AM is becoming a useful instrument for manufacturing planning, including design and fabrication of required production tools such as jigs and fixtures. Finally, AM affects the new product development process by enabling the use of a process with lower costs and which takes less time, in addition to having reduced design errors and quality costs. Studies demonstrate that AM also enables a shortening of the time to market. Reportedly, AM can cut new product development process costs by up to 70% and time to market by 90% (Waterman and Dickens 1994), and in addition it can reduce tooling lead times by 35% in comparison with conventional manufacturing (Morrow et al. 2007).

Exhibit 7—Impact on the New Product Development Process—Reducing the Time to Market
AM enables a rapid New Product Development (NPD) process through shortening the duration of one of the most important internally visible performance factors: the time to market. It enables rapid product launching from early initial idea to product completion.

Results of our survey show respondents strongly agree with the effectiveness of AM in reducing time to market (86.7%). In addition, results show that AM can considerably reduce the time needed to launch a new product. This capability has profound impacts on highly competitive markets. It is suggested, therefore, that companies in very competitive markets should implement AM in the processes of new product development and R&D. This results in reducing the costs of product development activities, decreasing risks of failure, and accordingly increases product competitiveness.

5.1.2   *Driving an Evolution in Design Methodologies*

Traditional design methodologies have four focuses; design for manufacturing (DFM), for assembly (DFA), for both (DFMA), and for disassembly (DFD). DFM guidelines encompass characteristics of the conventional manufacturing in order to have a good design. These guidelines help designers in developing modular designs, using standard components, designing multifunctional parts, minimizing assembly requirements, and avoiding imperfections during manufacturing such as varying wall thickness, and sharp corners. Moreover, they are focused on ease of assembly (DFA), and strategies that consider the future need to disassemble a product for repair, or refurbishment (DFD).

In fact, these design approaches are linked rather than being independent. Yang and Zhao (2015) compared these design approaches with respect to fourteen design attributes as follow:

(1) Design simply: complying with functional requirements
(2) Minimize part count
(3) Integrate parts
(4) Separate working components into modular subassemblies
(5) Minimize material types in an assembly
(6) Standardize components
(7) Create multi-functional parts
(8) Design for the ease of fabrication
(9) Design for the ease of assembly (positioning, handling, joining, and access)
(10) Avoid using laminates
(11) Avoid surface demands on components
(12) Avoid secondary operations
(13) Eliminate adjustments
(14) Use ferromagnetic materials.

Figure 5.2 illustrates the relationships between these four methods. A seen in the figure, items 1 (Design simply), 2 (Minimize part count), and 6 (Standardize components) are common to all, while the others have their own focuses. The point is that items 8 (Design for the ease of fabrication), and 9 (Design for the ease of assembly) are located on the periphery. This is due to the trade-off between DFM and DFA when considering design complexity and manufacturing constraints. However, thanks to AM technology this trade-off becomes redundant. In other words, the fundamental advantage of AM is the removal of this trade-off.

New design guidelines are needed to take into account the capabilities of AM technology. Design for Additive Manufacturing (DFAM) guidelines should involve

Fig. 5.2 Relationship between DFM, DFA, DFD, and DFMA. *Source* Yang and Zhao (2015)

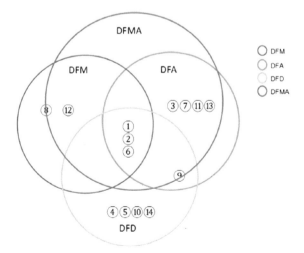

geometric possibilities and corresponding costs, developing integrated parts to reduce assembly and corresponding time, costs, and quality, reducing the weight of the part through use of porous structures, undercuts and hollow structures, thereby increasing functionality of the part through free from design.

Studies have attempted to develop design guidelines for AM technologies through empirical research. For instance, Becker et al. (2005) suggested some major design guidelines as follows:

- Use the advantages that are offered by AM processes
- Do not print the same parts as are designed for conventional manufacturing
- Do not consider conventional mechanical design principles
- Reduce the number of components by intelligent integration of functions
- Check if there are bionic instances to fit your tasks (these can give a suggestion towards a better design solution)
- Feel free to use freeform designs; they are no longer difficult to produce
- Optimize your design towards the highest strength-to-weight ratio
- Use undercuts and hollow structures if they are useful
- Do not worry about tooling as it is no longer needed.

5.1.3 Design Freedom

The toolless capability of AM empowers the production of parts with any geometry and complexity, a property known as freedom of design or freeform fabrication. Conventional manufacturing (namely the injection molding process) has some particular constraints for manufacturing that call for further considerations during the design process. These constraints result in a design approach called design for manufacturability (DFM). As stated in Sect. 5.1.2, this typology has to consider all manufacturing functions and assembly requirements during the design process.

From a technical point of view, AM—unlike injection molding—does not need subsequent melting within the mold. Thus, many corresponding problems such as inharmonic wall thickness, weld lines, sharp corners, sink and gate marks, and draft angles will not arise due to the toolless operations. Moreover, it is usually difficult to select the right location for a split line in the injection molding process. It is particularly difficult with parts that have a more complex shape, and it completely depends on the designer's experience. In addition, other design complexities may include blind holes, screws, and undercuts, etc. Although, it is not impossible to avoid these problems with conventional operations, layer-based manufacturing offers a much simpler and efficient solution (Mansour and Hague 2003).

Table 5.1 proposes a matrix in which a manufacturer can screen the processes and materials based on the given product's shape. This matrix compares the manufacturing processes (conventional and additive), and the available raw materials for use in a particular manufacturing method. In broad terms, the information demonstrates the flexibility of AM technologies to produce complex parts, but

Table 5.1 Process-material-shape matrix

Material				Shape / Processes	Circular Prismatic	Non-circular Prismatic	Flat Sheet	Dished Sheet	3D solid	3D hollow
Conventional processes										
Com	P	Cer	M	Sand Casting	✓	✓			✓	✓
			M	Die Casting	✓	✓			✓	✓
			M	Investment Casting	✓	✓			✓	✓
			M	Low pressure casting	✓	✓			✓	✓
			M	Forging	✓	✓				
			M	Extrusion	✓	✓				
			M	Sheet Forming	✓	✓	✓	✓		
		Cer		Powder Methods	✓	✓			✓	✓
		Cer		Electro-machining	✓	✓	✓		✓	✓
		Cer		Conventional Machining	✓	✓	✓		✓	✓
	P			Injection Molding	✓	✓	✓	✓	✓	✓
	P			Blow Molding						✓
	P			Compression Molding			✓	✓	✓	
	P			Rotation Molding				✓		✓
	P			Thermo-forming				✓		
	P			Polymer Casting	✓	✓			✓	✓
Com				Resin-transfer molding	✓	✓	✓	✓	✓	✓
Com				Filament Winding	✓	✓		✓		✓
Com				Lay-up methods			✓	✓		
Com				Vacuum Bag			✓	✓		
AM processes										
Com	P	Cer		Stereolithography (SLA)	✓	✓	✓	✓	✓	✓
Com	P	Cer	M	Multi-Jet Modeling (MJM)	✓	✓	✓	✓	✓	✓
Com	P	Cer	M	3D printing (3DP)	✓	✓	✓	✓	✓	✓
Com	P	Cer		Fused Deposition Modeling (FDM)	✓	✓	✓	✓	✓	✓
			M	Electron Beam Melting (EBM)	✓	✓	✓	✓	✓	✓
Com	P	Cer		Selective Laser Sintering (SLS)	✓	✓	✓	✓	✓	✓
			M	Direct Metal Laser Sintering (DMLS)	✓	✓	✓	✓	✓	✓
Com	P		M	Laminated Object Manufacturing (LOM)	✓	✓	✓	✓	✓	✓

Legend

M	Metals	Cer	Ceramics	Com	Composites	P	Polymers

Adapted from uz Zaman et al. (2017)

nevertheless, the range of raw materials is still limited with the available materials being metals, ceramics, composites and polymers. In addition, the shape criterion includes circular prismatic, non-circular prismatic, flat sheet, dished sheet, 3D solid and 3D hollows.

In other words, in conventional manufacturing the complexity of the product directly affects costs, while producing the same product with any level of complexity is independent of cost when using AM technology (Khorram Niaki and Nonino 2017b). Freedom of design also empowers the creation of integrated parts that do not need assembly, resulting in design optimization. This (re)design optimization, not only contributes to cost savings, but in many case also leads to higher reliability and product quality.

Furthermore, as mentioned in detail in Sect. 2.3.3, AM offers a unique capability in producing parts with an interior lattice structure to reduce weight. Lightweight parts are particularly beneficial for high-value products or raw materials. The increased geometrical complexity possible with AM facilitates an optimal design leading to a decrease in material consumption of up to 40% (Achillas et al. 2015). AM also allows for simple and cost-effective configuration in the design and redesign process (see Chap. 2 for practical examples). Therefore, AM allows designers to modify the design easily through many iterations, since the production of an early design is economically viable.

Another capability of AM is material combinations, whereas in the traditional casting process there was mostly one homogeneous material. AM potentially allows designers to put different materials into any required location of the part. This potential brings huge advantages to various industries. For instance, one of the potential future application of AM is in the production of smart textiles for smart wearable technology. There are generally two approaches for creating smart textiles for healthcare use. The first method is the creation of some form of conductive fibers, and the second is the addition of electronics to the surface. Using AM in this application sector not only enables one-step production in a single production volume, but also allows manufacturers to add (layer-by-layer) different materials and structures. This addition may include circuits, sensors, strain gauges, and energy harvester pieces. They may also add a short textile fiber on both sides, and materials like latex in the middle layers.

Given these effects together with the cutting of the tools and molds in the design process, restructuring of the design process and designer thinking is required because designers usually think of the traditional constraints in manufacturing, but now they have freedom of design—or at least new rules.

Exhibit 8—Impact on the New Product Development Process— Complexity for Free
Two important capabilities of AM distinguish it from other manufacturing methods. Firstly, AM allows for the production of highly complex parts. Because AM technology adds layer upon layer, it doesn't have the limitations of traditional methods such as molding and machining. It allow for the

printing of almost anything that the designer can draw. Thus, the production of complex or "impossible" geometries is now possible. This capability creates many possibilities for manufacturers in terms of designing and redesigning their existing products. Second, additively manufactured parts may have a better strength-to-weight ratio suitable to many industries such as the aerospace and automotive industries, in which the weight of the final product is very important (see Chaps. 2 and 3 for more detail). For instance, AM enables the direct fabrication of lattice structures with gradual and controlled porosity in order to reduce weight. Moreover, for products with high value raw materials, it can also be very beneficial as it results in the use of less material. In other words, AM results in a reduction of a product's weight while increasing the strength-to-weight ratio that is the key success factor in many manufacturing environments.

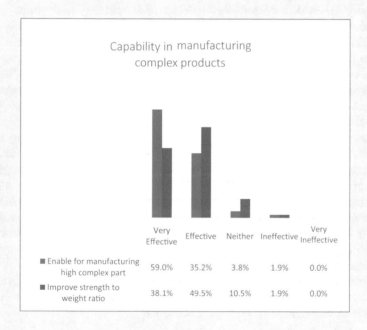

More than 94% of respondents identified the effectiveness of AM in making products with a great shape and dimensional complexity. Thus, industries that need to redesign their existing products are highly encouraged to implement AM technology. The redesigned part can, for instance, be an integrated part instead of many components that need to be assembled. Moreover, more than 87% of respondents affirmed the effectiveness of AM in making lightweight products with higher strength-to-weight ratios. This enables many industries to save on the costs of materials as well as making product lifecycle costs.

5.2   The Value for Manufacturing

Additive technologies present great opportunities for current and future manufacturing. Their impact on production processes, product quality, and manufacturing costs is significant. The implementation of AM will affect Just-In-Time (JIT) manufacturers through:

- *Dematerializing Supply Chain*—Since AM only needs a 3D CAD design to initiate production, it results in a very simple supply network, called a dematerialized supply chain.
- *Just-In-Time manufacturing*—The application of AM in the manufacturing environment will reduce material distribution and inventory costs for work-in-progress (WIP). In other words, all costs associated with the WIP can be removed through the use of AM, so that AM controls JIT manufacture at the factory, rather than the traditional concept of JIT delivery to the firm.
- *Reducing setup and changeover time and cost, and number of assemblies*—AM machines do not need tooling for the production of different parts, therefore, in comparison with conventional manufacturing, AM is in the lead, reducing setup and changeover times and costs. In addition, manufacturers have traditionally preferred to have integrated parts with reduced assembly requirements, to improve the part's functionalities and reduce operational cost. This is already more feasible with AM technologies.
- *Reducing waste*—The principle objective of the lean paradigm is to reduce waste throughout the supply chain. As 3D CAD data is one of the manufacturing resources, the data can be directly sent to AM machine from any designer or supplier through the use of internet technologies. This results in a faster exchange of data between design suppliers and manufacturers. Moreover, AM enables the economical production of single parts, thus the holding costs that were traditionally necessary can be eliminated or at least reduced with AM. Additionally, AM enables a shift in production location, so that retailers, assembly lines and customers can be closer to consumers. This results in a reduction of logistic costs, relating to both internal and external logistics. In summary, it can be concluded that AM will enable manufacturers to reduce waste in materials, time, cost and distribution.

5.2.1   Impact on Production Process

Production planning and control (PPC) refers to the activities that make the production and delivery of products flatter and more efficient in terms of time and cost. PPC will become less complex thank to AM. In these systems, in which products are built through CAD data, most of the information required for production planning can be sent automatically to an AM machine.

Fig. 5.3 A CAD model on the left converted into STL format on the right. *Image source* Gibson et al. (2010)

The first input in production is the 3D CAD data of the object or a scan of the physical parts. This step is also known as computer-aided engineering (CAE). Then, specialized software translates the data to a suitable file format (namely STL), readable by an AM machine. STL is a generic format, developed by 3D Systems, to translate CAD data into information for each printed layer (an example of which is shown in Fig. 5.3.

This software slices the design into layers, in which the new file contains information on each layer. In CAE, process skills optimization refers to the adaption of the part to the manufacturing technology chosen. The schematic of the AM process is illustrated in Fig. 5.4. During the design process, designers should also include supplementary steps such as support structure design, and selecting process parameters (e.g. build orientation), which is shown as a configuration step in Fig. 5.4.

Exhibit 9—Impact on Manufacturing Process—Reducing the Manufacturing Time?

Production time refers to the period from the start of the print to the finishing and post-processing steps of a part. Generally, AM is more time consuming on a per part basis than any other conventional mass production method. However, in some cases it can be considered as the faster manufacturing method, particularly when the component needs to be customized or self-fabricated. Since AM enables the creation of functional parts without the need for assembly, it may result in savings in both production time and cost. Moreover, since the production time is based on the number of layers, it is related to product size. It means that a larger product size needs more printing time than a smaller one.

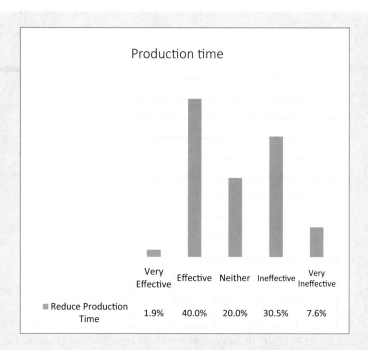

Manufacturing time is the sum of time spent for:

1. processing material (run time)
2. materials waiting to be processed (queue time)
3. setting machinery up for the next operation (set-up time)
4. moving material from one processing stage to another (move time).

AM is still not still an effective technology for reducing production time. In fact, more than 58% of respondents believe that AM needs more process time in comparison with conventional manufacturing. This demonstrates that AM machinery has to invest more in improving the performance of production time, so that it becomes competitive with traditional manufacturing methods.

5.2.2 Impact on Product Quality

Quality means satisfying customer needs whilst providing superior value. Empirical studies have revealed that the current technological shortcomings are poor dimensional accuracy and surface finish. This depends on several parameters such as part geometry, type of AM system, material properties, post-processing, and

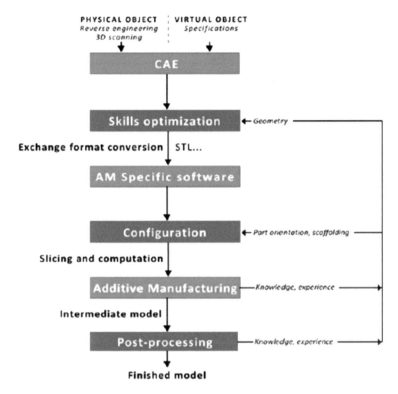

Fig. 5.4 AM production process. *Source* Gardan (2016)

intermediate steps. For instance, the main disadvantages of the FDM system are surface roughness with a grainy appearance, and relatively poor dimensional precision.

AM technology is unable to manufacture parts that require high accuracy of dimensional measurements. Table 5.2 compares the dimension tolerance ranges of conventional processes and additive processes. Although AM technology might not be as good as traditional machining processes (such as subtractive techniques like milling), it is suitable for final product quality when compared with conventional cast-molding processes. As shown in exhibit 10, the quality performance of the final 3D printed parts is not as good as that in conventional manufacturing.

Table 5.2 List of dimensional tolerances for CM and AM processes

	Manufacturing process	Tolerance (mm)
Conventional manufacturing	Sand casting	1.00–5.50
	Die casting	0.26–0.71
	Investment casting	0.1–0.59
	Low pressure casting	0.23–0.59
	Forging	0.59–3.22
	Extrusion	0.34–3.22
	Sheet forming	0.10–3.22
	Powder methods	0.59–1.30
	Electro-machining	0.03–0.10
	Conventional machining	0.01–1.00
	Injection molding	0.10–1.00
	Blow molding	0.43–1.00
	Compression molding	0.26–1.04
	Rotational molding	0.60–1.00
	Thermo-forming	0.35–1.00
	Polymer casting	0.79–2.20
	Resin-transfer molding	0.36–1.00
	Filament winding	0.66–2.20
	Lay-up methods	0.70–2.30
	Vacuum bag	0.66–2.20
Additive manufacturing	Stereolithography (SLA)	0.025–0.10
	Multi-jet modeling (MJM)	0.10–0.20
	3D printing (3DP)	0.20–0.40
	Fused deposition modeling (FDM)	0.05–0.5
	Electron beam melting (EBM)	0.20–0.40
	Selective laser sintering (SLS)	0.1–0.46
	Selective laser melting (SLM)	0.05–0.10
	Direct metal laser sintering (DMLS)	0.10–0.20
	Laminated object manufacturing (LOM)	0.25–0.50

Adapted from Uz Zaman et al. (2017)

Exhibit 10—Impact on the Manufacturing Process—Improving Product Quality?
Currently, studies and practitioners confirm that the main technological shortcomings include the relatively poor dimensional accuracy and the surface finish.

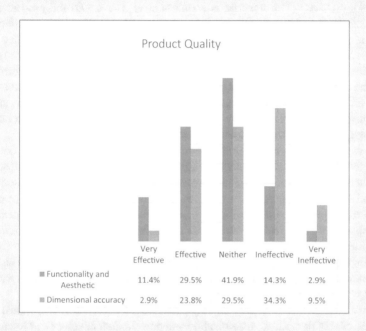

	Very Effective	Effective	Neither	Ineffective	Very Ineffective
Functionality and Aesthetic	11.4%	29.5%	41.9%	14.3%	2.9%
Dimensional accuracy	2.9%	23.8%	29.5%	34.3%	9.5%

The results show that 59.1% of respondents do not believe in the effectiveness of AM technology regarding functionality and aesthetics, and more than 73% feel the same regarding dimensional accuracy. Therefore, the current use of AM is not suitable for producing high quality precision parts. However, in recent years many technological advancements have been made in AM machinery, and research shows that in the future machines will be more accurate. In addition, in some cases, the use of a traditional machining process for 3D printed parts may improve the final product.

5.2.3   Impact on Manufacturing Costs

Generally, AM production costs consist of various components such as equipment, labor, materials, and pre- and post-processing. However, AM generates a shift in production cost arrangements towards a high share (45–75%) of machinery costs in

the total production costs, depending on the individual case. In a production case
Lindemann et al. (2012) showed that raw material costs are estimated at (only) 12%
of the total AM production costs.

This case study, in which the costs of the AM process were clearly divided, will
now be looked at in greater depth. Figure 5.5 shows the sample part (on the left
side) that is an automotive component that should be produced using an AM
technology (SLM). The material is stainless steel 316 L and the process rate is
6.3 cm^3/h. The cost breakdown then shows the share of each of the manufacturing
steps respectively: machine costs are 73% of the total costs, materials 12%,
post-processing 7%, oven 1%, building-process fix 3%, and costs for preparation
4%. The machine costs were calculated based on the following information:
machine utilization: 4500 h/year; machine depreciation time: five years; investment
costs: €500,000; and costs for maintenance €21,666/Year. Prices for AM materials
are considerably higher than raw materials for conventional processes, in addition
to the need for amortization due to higher material efficiencies.

These technologies can currently be considered as both feasible and economical
alternatives to conventional manufacturing due to the phenomenal impacts on
manufacturing and product design. However, it can now compete in the production
of small and medium production volumes (Lot-Size). As stated by practitioners, the
machine cost per part still contributes the major part of the total cost. As mentioned
in Chap. 4, technological advancements are going to mean cheaper resources for
AM in terms of both materials and equipment as we have seen in the growth in
manufacturing of affordable 3D printers in recent years. Likewise, AM is expected
to become increasingly cost effective for larger products and production volumes
than it is at present.

Of course AM allows a company to benefit from a decrease in operational costs:
from the cost of flexibility, customization and new product development to marginal

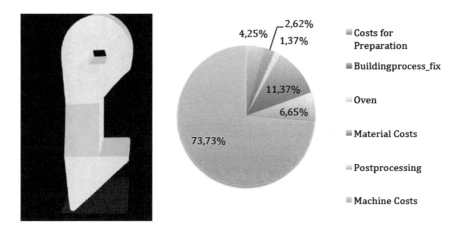

Fig. 5.5 Sample part, and cost breakdown for a production volume of 190 units. *Source*
Lindemann et al. (2012)

production costs. Nevertheless, as seen in Sect. 3.1.1, when production volume increases AM becomes less competitive compared to manufacturing based on conventional FMS, where costs per unit decrease as volumes increase. Accordingly, the smaller the production volume, the more competitive AM becomes when the costs per unit of conventional FMS increase. Moreover, the incentives to adopt AM technologies are few when the target performances for time and quality are too high.

Exhibit 11—Impact on Manufacturing Process—Reducing the Labor Cost
As regards workforce, because AM technologies do not need a multifunctional team for the design and running of the production line—as in most of the conventional methods—it results in a lower labor cost. In addition, an operator can simultaneously run and control multiple machines. Moreover, thanks to the newly advanced combination of AM with robots, greater process automation has been achieved, in which robots plays the role of operators feeding the AM machine. It can also work during the night without the need for human presence. However, in several AM-based machines, post-processing still needs to be done by a person.

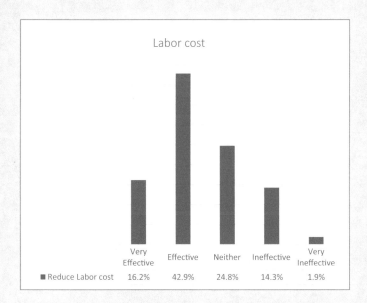

Approximately 59% of the respondents agree with the effectiveness of AM in reducing labor costs, however, a higher percentage was expected. This performance indicator can be improved by further attempts for automated process, and reducing the need for post-processing.

5.2.4 *Impact on Material Waste*

AM uses only the material necessary to create parts, leading to a minimal amount of waste compared to traditional machining that cuts the bulk of raw material required. In addition, it offers not only minimal wastage in terms of raw materials, but also enables a higher recycling rate of raw materials compared to traditional machining or casting methods. This is demonstrated by studies showing that the waste material in metal applications associated with AM is reduced by 40%, while 95–98% of waste material can be recycled (Petrovic et al. 2011). Consequently, reducing material waste and a higher recycling rate reduces costs. The only waste is the support structure that is being used in some AM systems, and this too may be recyclable. Moreover, it is obvious that the waste reduction offered by AM contributes to the saving of energy during production and other related sources such as water and fuel consumption.

Exhibit 12—Impact on Manufacturing Process—Reducing the Material Waste
Studies demonstrate that AM technologies can use minimal raw materials in comparison with other manufacturing technologies for a given part.

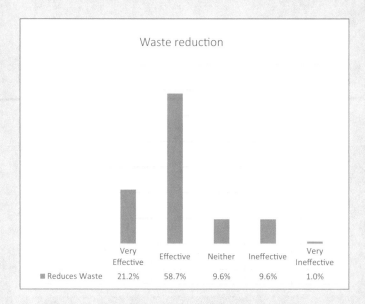

The capability of AM in reducing waste (in both general efficiency and material usage) produced higher scores. About 80% of the surveyed companies recognized the effectiveness of AM in reducing material usage. Therefore, it is highly recommended to implement AM technologies in the case of high-value raw materials.

5.3 The Value for Logistics and Supply Chain Management

AM provides a number of advantages, such as production locations that are closer to end consumers, a dematerialized supply chain, a high level of customization, low production lead times, lower logistics costs, and minimal inventories (Liu et al. 2014; Tuck et al. 2007). Due to these significant impacts on the supply chain, several studies have been carried out in this area.

To understand the real effects of AM on the supply chain process, Oettmeier and Hofmann (2016) conducted an empirical study. This case study involved two companies (one large and one SME), both in the hearing aid industry, which needs to produce highly customized products. These companies transitioned to AM from their previous manufacturing method of handcrafting. Both companies have realized many advantages in supply chain processes and components after adopting AM.

In terms of supplier relationships, one company that has many production locations worldwide could standardize logistic operations, in which the raw material, equipment and production requirements are the same at all production locations, leading to greater control over product quality. As for the manufacturing flow, after adopting AM both companies now produce customized parts in a production volume of twelve to forty in a single run, whereas formerly, when using manual manufacturing they produced only a single part and batch production was not possible. However, after employing AM, they needed more skilled labor (namely one person for 3D modelling, one for operating the AM machine, one specialist for testing parts, where previously a single person carried out the manufacturing. This means more industrialization of the manufacturing process and employees with special tasks and training requirements, whereas formerly only one person was responsible for all manufacturing steps. In fact, in comparison with producing customized parts through manual operation, AM has several advantages such as industrialization of fabrication, a new operational organization with specific tasks and training requirements, and corresponding quality matters. In addition, the transition from manual operations to AM has fostered the generation of economies of scale through batch production.

In the product development process and commercialization, they found that the freedom of design offered by AM to be a significant effect as they are now able to produce a shell with a smaller and controllable thickness, which allows more functionality to be added. However, according to informants, more education and training programs are needed so that more can be achieved from the potential that freedom of design promises. Also, thanks to rapid prototyping, customers can be integrated into the product development process earlier, leading to greater market acceptance. As for customer relationships, one of the companies has succeed in digitalizing its connections with customers in order to fulfill their orders. In this system, an acoustician scans the ear impression and sends it digitally to the company, eliminating physical delivery and leading to faster initiation of production.

Finally, the return management process profits from an increasing material utilization rate. Reportedly, thanks to AM, the material utilization rate of the companies has increased from 98 to 100%, because in the AM process uncured material can be reused, while in conventional casting the excess material was typically disposed of. However, specific reprocessing of unused raw materials may be required (e.g. filtering) to maintain the quality of the AM materials. Furthermore, in comparison with the previous manual manufacturing method, AM accelerates the displacement process and improves replicability due to the digitized process and stored 3D model data.

5.3.1 The Impact on Inventory Turnover

Using AM for producing parts on-demand can definitely slash the costs of holding and inventories, for both final products and work in process. AM can be responsible for a move from a centralized supply chain to distributed and local manufacturing (see Sect. 5.3.2 for details), where the first approach requires a higher safety inventory in order to satisfy the broad range of customers, while local manufacturing allows for a minimum amount of inventory and logistic costs. Moreover, using AM technologies results in so-called dematerialized supply chains. Conventional manufacturing requires several raw material, parts and tools to produce a product, while AM usually uses only a single material plus a CAD design as manufacturing resources. In addition, since AM allows for the production of an integrated and more functional product, there is no need to hold different parts that need to be assembled. This flexible manufacturing system will have a profound impact on highly innovative products with fluctuating demand rates in particular. Also, with respect to a high value good, even though conventional manufacturing would have a lower operational cost, the higher inventory cost forces the following of a make-to-order approach.

> **Exhibit 13—Impact on Supply Chain Management and Logistics— Reducing the Inventory Turnover**
> AM empowers companies to reduce the cost of inventory mostly through make-to-order production. Inventory turnover can be minimized through the dematerialized nature of the AM supply chain, and the closer production location of manufacturers to end consumers. This factor offers production flexibility and capital efficiency that contributes to the lean paradigm, and so eliminates almost all processes that do not participate in the creation of added value. Therefore, it can be seen as a useful method for products with fluctuating demand.

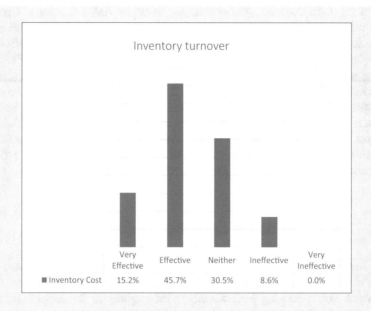

60.9% of the companies surveyed recognized the effectiveness of AM in reducing inventory turnover. This is the second highly scored cost performance that is related to logistics and SCM (see Table 5.3).

Table 5.3 Operational performance based on the operation strategies and processes

OPERATION PROCESS / OPERATION STRATEGY	NPD process		Manufacturing Process		SC & Logistics	
Time	Reducing Time to Market	87%*	Production Time	42%	-	
Cost	-		Labor Cost	59%	Reducing Inventory Turnover	61%
	-		Waste reduction	80%	-	
Quality	Complexity for free	94%	Product Aesthetic	40%	-	
	Strength-to-weight ratio	87%	Dimensional Accuracy	27%	-	

* Percentage of respondents that recognized the effectiveness of AM technology

Exhibits 7–13 discussed the operational value of AM technology. Operation performance is classified in terms of time, cost and quality. In addition, this performance deals with either the new product development process (NPD), manufacturing, or supply chain and logistics processes. Table 5.3 presents an overview of surveyed operational performance and the percentage of respondents that recognized the effectiveness of AM technology relating to the subject in question. As can be seen in the table, on average, the best performances occur in NPD and supply chain processes. In terms of operation strategies and performance, factors dealing with cost are the most advantageous. In total, the best performances are complexity for free and strength-to-weight ratio. The worst performances are, respectively, a product's dimensional accuracy, functionality and aesthetic, and production time.

5.3.2 The Impact on the Spare Part Supply Chain

The spare part supply chain has usually been a complicated process because of the role of slowly moving parts and an unpredictable demand pattern. Typically, a small number of slowly moving parts contribute the majority of the supply chain cost. Conventional manufacturing technology requires a large batch size to be produced in order for it to take advantage of economies of scale. For a slowly moving part, a large batch size is not economically justifiable while also causing an additional inventory cost, however, AM enables the reduction in operational cost for a small production volume. Additionally, companies should hold a variety of spare parts for a long period, and in a high number—even of some out of date goods for after-sale services. Together with the transportation cost, it results in stagnant capital.

AM enables a decreased stock level and a corresponding obsolescence risk to parts because of the possibility of make-to-order production. Furthermore, for big companies with worldwide business markets, spare part transportation makes up a big proportion of supply chain costs, which can be reduced with distributed manufacturing and printing at the consumer's location. This is also leading to faster supply times and maximum responsiveness.

AM also brings many opportunities in the use-phase of spare parts. For instance, it can provide a longer usage period with a lower cost, through the possible repairing of worn out parts, leading to substantial cost savings. Another capability of AM may be the producing of a part that is more reliable so as to increase the replacement interval, resulting in a reduction in total lifecycle costs. It is also possible for the user to create a temporary fix in the case of the replaceable part being unavailable. The newly printed part would be useful until the projected replacement becomes available. However, although the printed part might have a lower performance, it can provide an alternative to storing some expensive parts.

Generally, Table 5.4 summarized the different levels of impact that the implementing of AM may offer to companies, given the fundamental attributes of the spare part supply chain. These attributes include demand rate, resupply lead time, agreed response time, remaining usage period, manufacturing/order cost, safety stock cost, number of supply options, and supply risk.

Table 5.4 Influence of AM technologies on spare part supply chain attributes

Spare part attributes	Potential impacts						
	Reduce manufacturing/order cost	Reduce direct part usage cost	Reduce safety stock cost	Improve supply chain responsiveness	Postponement	Temporary fix	Reduce effect of supply
Demand rate	Low		Low		Low		
Resupply lead time			Long	Long	Long	Long	
Agreed response time			Short	Short		Short	
Remaining usage period		Long					
Manufacturing/order cost	High						
Safety stock cost			High		High		
Number of supply options	Few			Few			Few
Supply risk				High			High

Sources Knofius et al. (2016)

Within the problem of the spare part supply chain and the impacts of AM on it, the aerospace industry was the point of interest for several researchers (Walter et al. 2004; Hasan and Rennie 2008; Holmström et al. 2010; and Liu et al. 2014), because of the higher value of the aircraft spare parts, the slow moving parts and the time-consuming manufacturing of parts. The demand in this industry usually follows the 20/80 Pareto curve, which means that 80% of the spare parts are needed frequently, however, it only accounts for 20% of the supply chain cost in terms of inventory. In fact, the other 20% of spare parts are needed infrequently, however, the majority of the supply chain expenditure is due to this 20% of slowly moving parts. In addition, the manufacturing lead time is usually very long when using such traditional manufacturing techniques, and this industry requires a very high cycle service level to satisfy demands. This leads to the maintenance of a further level of safety inventory, resulting in additional costs.

A study by Liu et al. (2014) evaluated the impact of AM in the aircraft spare part supply chain based on the Supply Chain Operation Reference (SCOR) model. They investigated three supply chain scenarios, namely conventional (as-is), centralized AM, and distributed supply chain. Firstly, it is necessary to briefly define the SCOR model. The SCOR framework includes a set of evaluation measures of business processes, technology and best practices due to the effective exchange of materials and information between different levels of supply chain. It identifies five key performance indicators of supply chains such as reliability, responsiveness, agility, costs and asset management.

As mentioned above, three scenarios have been evaluated in order to adapt AM technologies to the aircraft spare part supply chain. First, is the conventional "as-is" in which the original equipment manufacturers (OEM) produce the parts and supply them to the distribution centers, which in turn supply parts to different service locations (see Fig. 5.6a). Service locations are the place of maintenance, repair or overhaul. In the centralized scenario, spare parts are manufactured in a regional distribution center by AM technologies (see Fig. 5.6b), which offer efficient economies of scale to meet the demands of service locations. The operation in regional distribution centers follows the make-to-order pattern, however, a safety inventory is still required in order to preserve a high-cycle service level. The advantages of the implementation of AM in this system include a shorter manufacturing lead time in comparison with some conventional methods, a shorter logistic lead time because the distribution centers are closer to the consumers, and a reduced amount of safety inventories. In the distributed scenario, spare parts are manufactured in service locations rather than distribution centers by AM technologies (see Fig. 5.6c), resulting in lower logistic and inventory costs compared with the previous two scenarios.

Using the SCOR model measurements, the results of case studies revealed interesting conclusions. Introducing AM to an aircraft spare part supply chain definitely improves the conventional configuration in terms of required safety inventory and inventory cost of the whole supply chain. It is also demonstrated that the centralized AM supply chain is preferable for lower demand parts, high fluctuations in demand, and a longer manufacturing lead time. By contrast, the distributed AM supply chain is more suitable for parts with a high average demand,

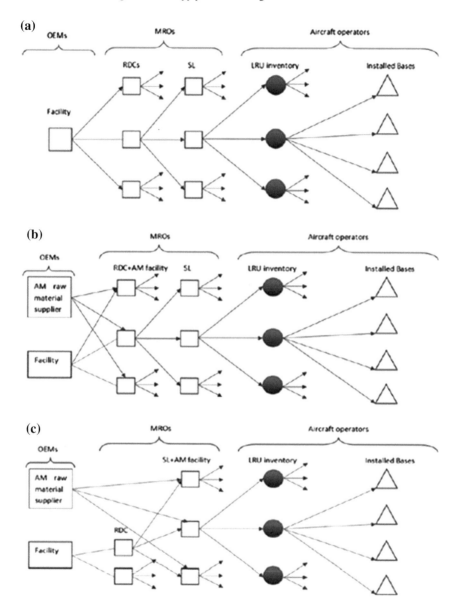

Fig. 5.6 Aircraft spare parts supply chain configurations: **a** as-is, **b** centralized, **c** distributed supply chain. *Source* Liu et al. (2014)

and a very stable demand. In this case, implementation of AM and its corresponding investments need a return on investment (ROI) analysis as the high cost of the initial investments might not be returned quickly.

Fig. 5.7 Four scenarios
based on the status of
development of AM machines
and supply chain
configurations. *Source*
Khajavi et al. (2014)

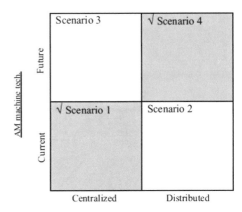

To analyze the costs of a distributed spare part supply chain, Khajavi et al. (2014) conducted further empirical investigations on the procurement of F/A-18E/F Super Hornet fighter jets (for the US navy). In their analysis, aside from the supply chain configuration (distributed and centralized) is another factor—the development of AM machines—which also has to be considered. The development of machines refers to the expected advancements of AM machines in the future, which may lead to further reductions in overall cost throughout the supply chain, and further improvements in productivity. These advances may be a smaller build chamber that lowers the price of the AM machine, highly automated processing (especially in pre- and post-processing), and a higher production speed.

Considering these aspects (as seen in Fig. 5.7), four scenarios are possible. For instance, using current AM machines in a distributed supply chain (scenario 2) or employing future machines in a centralized configuration (scenario 3). To analyze these four supply chain strategies, the total costs of the whole supply chain were compared. The major cost components include workforce cost (as it varies in different scenarios—in a distributed supply chain it should be more than a centralized one, and in future AM machines it is expected to be less), transportation (less cost in a distributed supply chain than a centralized one), inventory turnover (different stock and safety inventory level in supply chain configurations), initial investment (which depends on both supply chain configurations and machine development level—future machines are expected to need less initial investment).

As seen in Fig. 5.7, the study of Khajavi et al. (2014) concluded that centralized supply chain is more efficient for the current AM machines, while the future advances in AM machines mean distributed supply chain will be feasible. Using current AM machines, centralized production results in lower operational costs than distributed manufacturing. This is caused by the current high cost of equipment, together with the need to purchase more machines in a distributed supply chain. Consequently, the higher number of machines in distributed supply chain also needs more labors. Although distributed manufacturing results in a lower inventory turnover and transportation cost, when taking into account the current specification of AM machines, it doesn't seem so efficient. To take advantage of distributed manufacturing, AM machines need to be developed further.

These technological advances should result in a smaller build chamber, a cheaper machine and a more automated process, in which a distributed production (scenario 4) may be viable. This is very important in supply chain configurations and can be seen as a radical change not only in spare part supply chains, but it also has fundamental advantages in the supply chains of goods produce by a distributed manufacturing approach. These advantages include lowering operational costs, accelerating demand fulfillment, improving customer relationships, faster after-sale service, fewer environmental impacts due to a reduced need for transportation, flows of stagnant capital (due to lower stock levels and inventory turnover), geopolitical advantages and so on.

5.3.3 3DP Online Platforms Supply Chain

Figure 5.8 illustrates the supply chain (both product and information flows) of 3DP online platforms in which generative services include those related to design file preparations. Facilitative services include uploading and storing data as well as customizing (redesigning) activities based on the requirements of the customers. Selective services involve all activities related to offering 3D models, from which customers can choose. After preparing the 3D model data, the next stage is the transfer of data to the manufacturing stage and fabrication of the object. The final stage is distribution and this is usually handled at the manufacturing facility by packaging, and then by a third-party transportation service or the company's local retailer.

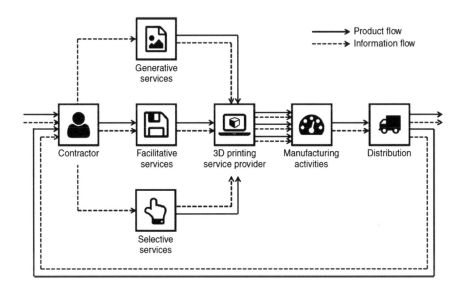

Fig. 5.8 The supply chain of 3D printing manufacturing platforms. *Source* Rogers et al. (2016)

Various web-based frameworks for these 3D service platforms have been developed. Lan et al. (2005) led AM in a networked manufacturing system structure by taking full advantage of the rapidly evolving computer network and information technologies. In their work, some key modules are involved, including modelling and planning a manufacturing chain, selecting feasible collaborative manufacturers, queueing a manufacturing task, using the synchronously collaborative work environment, and constructing a suitable running platform. Furthermore, it may involve collaborative product commerce (CPC), which includes collaborative planning and control of production, concurrent process of RP and RM, optimization of external manufacturing resources, online services to customers, and detailed structure and formulation of the central-monitoring mechanism of such a partnership system. Networked manufacturing based on the web is a new manufacturing method in terms of the mission, structure, infrastructure, capabilities, and design process.

Tele-manufacturing facility (TMF) for instance should enable users to directly access AM facilities through the internet. In these systems, users can submit their jobs and be in the queue for production. Figure 5.9 illustrates the overall architecture of a typical tele-manufacturing service for rapid prototyping. The system has been established and employed in the Northwest Productivity Promotion Center in China. These systems allow different enterprises—from anywhere—to share their 3D printing machines and services. It includes two main components—a software sub-system and a hardware sub-system. The former involves eight components including information center, application service providers (ASP) tool set, client management, e-commerce, system navigation, manufacturing service, and collaborative tools. The latter (hardware) includes not only the AM equipment of the bureaus but also the AM machines of other service bureaus. These bureaus are equiped with different units including CAD work stations, reverse engineering (RE) tools, and AM equipment (such as for rapid prototyping and rapid tooling). Some special modules for the introduction of AM are discussed in the following paragraphs, however, some of them are common to designing a framework for web-based business in any circumstances.

The ASP tools consist of five components, namely STL file checking and fixing, support structure generation, process planning and optimization of process parameters, AM technology selectors, and build time estimation. Once, a quote is delivered, those packages start to analyze the given product specifications and initiate production planning. It provides suitable and best-fit AM technologies for the specific requested part. The system then can automatically check for flaws in the STL (STereoLithography) files. The process-planning stage determines tool paths and process parameters, which may include part orientation, slicing, support structure generation, and path planning.

Electronic commerce unit involves four components—the online quote system, build-time estimation (for price and delivery estimation), online business negotiation, and electronic contract management. At the time that clients initially accept the quote, they may negotiate with the supplier on the business and technological particulars. This facility is provided by, for instance, Microsoft NetMeeting, which offers a collaborative environment in which to share information, and transfer files,

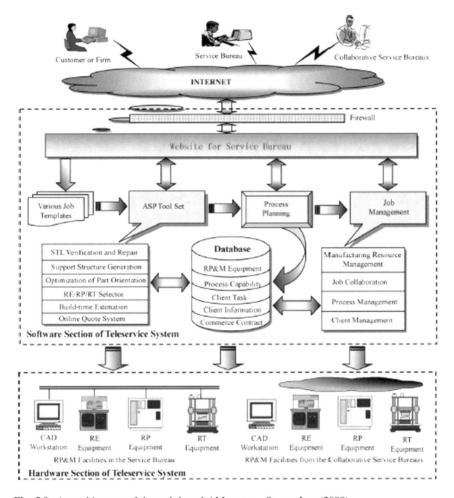

Fig. 5.9 An architecture of the web-based AM system. *Source* Lan (2009)

video or audio communication etc. After these negotiations, a contract can be electronically signed, and this is included in the system.

The manufacturing service components, which involves job management, process monitoring, collaborative manufacturing and enterprises management, is one of the important functional modules of the system. Once the contract is signed, a client provides their requirements and sources (e.g., 3D model or STL file). Collaborative manufacturing determines the collaborative enterprise to assign the job to. In addition, it then monitors the process and its schedule in order to ensure correct and efficient production.

Seven Key Facts

- AM can change the new product development process through reducing cost, time, design errors, and the cost of quality.
- Design for Additive Manufacturing (DFAM) has been developed to take into account the capabilities of AM technology and render conventional constraints redundant.
- The ability to produce highly complex, lightweight parts distinguishes AM technology from other manufacturing methods.
- AM enables manufacturers to reduce waste in materials, time, cost and distribution.
- AM is not still considered as an effective technology for reducing production time.
- AM does not need multifunctional teams for design and running of the production line, as most conventional methods do, thus it has a lower labor cost.
- AM slashes the costs of holding and inventories, both for final products and the work in process inventory.

References

Achillas, C., Aidonis, D., Iakovou, E., Thymianidis, M., & Tzetzis, D. (2015). A methodological framework for the inclusion of modern additive manufacturing into the production portfolio of a focused factory. *Journal of Manufacturing Systems, 37,* 328–339.

Becker, R., Grzesiak, A., & Henning, A. (2005). Rethink assembly design. *Assembly Automation, 25*(4), 262–266.

De Beer, D. J., Campbell, R. I., Truscott, M., Barnard, L. J., & Booysen, G. J. (2009). Client-centered design evolution via functional prototyping. *International Journal of Product Development, 8*(1), 22–41.

Gardan, J. (2016). Additive manufacturing technologies: State of the art and trends. *International Journal of Production Research, 54*(10), 3118–3132.

Gibson, I., Rosen, D. W., & Stucker, B. (2010). *Additive manufacturing technologies.* New York: Springer.

Hasan, S., Rennie, A. E. W. (2008). The application of rapid manufacturing technologies in the spare parts industry. In *Nineteenth Annual International Solid Freeform Fabrication (SFF) Symposium, 4–8 August 2008,* Austin, TX, USA.

Holmström, J., Partanen, J., Tuomi, J., & Walter, M. (2010). Rapid manufacturing in the spare parts supply chain: Alternative approaches to capacity deployment. *Journal of Manufacturing Technology Management, 21*(6), 687–697.

Khajavi, S. H., Partanen, J., & Holmström, J. (2014). Additive manufacturing in the spare parts supply chain. *Computers in Industry, 65*(1), 50–63.

Khorram Niaki, M., & Nonino, F. (2017a). Additive manufacturing management: A review and future research agenda. *International Journal of Production Research, 55*(5), 1419–1439.

Khorram Niaki, M., & Nonino, F. (2017b). Impact of additive manufacturing on business competitiveness: A multiple case study. *Journal of Manufacturing Technology Management, 28*(1), 56–74.

Knofius, N., Knofius, N., van der Heijden, M. C., van der Heijden, M. C., Zijm, W. H. M., & Zijm, W. H. M. (2016). Selecting parts for additive manufacturing in service logistics. *Journal of Manufacturing Technology Management, 27*(7), 915–931.

Lan, H., Ding, Y., Hong, J. (2005). Decision support system for rapid prototyping process selection through integration of fuzzy synthetic evaluation and an expert system. *International Journal of Production Research, 43*(1), 169–194.

Lan, H. (2009). Web-based rapid prototyping and manufacturing systems: A review. *Computers in Industry, 60*(9), 643–656.

Lindemann, C., Jahnke, U., Moi, M., Koch, R. (2012, August). Analyzing product lifecycle costs for a better understanding of cost drivers in additive manufacturing. In *23rd Annual International Solid Freeform Fabrication Symposium–An Additive Manufacturing Conference.* Austin, TX, USA 6th–8th August.

Liu, P., Huang, S. H., Mokasdar, A., Zhou, H., & Hou, L. (2014). The impact of additive manufacturing in the aircraft spare parts supply chain: Supply chain operation reference (SCOR) model based analysis. *Production Planning & Control, 25*(13–14), 1169–1181.

Mansour, S., & Hague, R. (2003). Impact of rapid manufacturing on design for manufacture for injection molding. *Proceedings of the Institution of Mechanical Engineers, Part B: Journal of Engineering Manufacture, 217*(4), 453–461.

Morrow, W. R., Qi, H., Kim, I., Mazumder, J., & Skerlos, S. J. (2007). Environmental aspects of laser-based and conventional tool and die manufacturing. *Journal of Cleaner Production, 15* (10), 932–943.

Oettmeier, K., & Hofmann, E. (2016). Impact of additive manufacturing technology adoption on supply chain management processes and components. *Journal of Manufacturing Technology Management, 27*(7), 944–968.

Petrovic, V., Vicente Haro Gonzalez, J., Jorda Ferrando, O., Delgado Gordillo, J., Ramon Blasco Puchades, J., Portoles Grinan, L. (2011). Additive layered manufacturing: Sectors of industrial application shown through case studies. *International Journal of Production Research, 49*(4), 1061–1079.

Rogers, H., Baricz, N., & Pawar, K. S. (2016). 3D printing services: Classification, supply chain implications and research agenda. *International Journal of Physical Distribution & Logistics Management, 46*(10), 886–907.

Tuck, C., Hague, R., & Burns, N. (2007). Rapid manufacturing: Impact on supply chain methodologies and practice. *International Journal of Services and Operations Management, 3* (1), 1–22.

uz Zaman, U. K., Siadat, A., Rivette, M., Baqai, A. A., & Qiao, L. (2017). Integrated product-process design to suggest appropriate manufacturing technology: A review. *The International Journal of Advanced Manufacturing Technology, 91*(1–4), 1409–1430.

Walter, M., Holmström, J., Tuomi, H., Yrjölä, H. (2004, September). Rapid manufacturing and its impact on supply chain management. In *Proceedings of the Logistics Research Network Annual Conference* (pp. 9–10).

Waterman, N. A., & Dickens, P. (1994). Rapid product development in the USA, Europe and Japan. *World Class Design to Manufacture, 1*(3), 27–36.

Yang, S., & Zhao, Y. (2015). Additive manufacturing-enabled design theory and methodology: A critical review. In *International Journal of Advanced Manufacturing Technology, 80.*

Chapter 6
Strategic Alignment of Additive Manufacturing

Previous chapters underlined the value of AM to industry, business and operation strategies, and manufacturing and operations. Regarding the general impacts, our survey results (illustrated in the exhibits) identify and confirm the positive impacts on time, cost, quality, environment and business flexibility performance. It is not feasible to identify a specific factor common to all circumstances, or to generalize all the factors to a specific circumstance. The chapter proposes a framework containing some of the building blocks necessary for the strategic alignment of a company's business model with the adoption of AM. It discusses business and operations strategies for introducing additive technologies and continues with identifying contingency factors driving AM performance. These factors are categorized as organizational, operational, and products' characteristics. The chapter ends with an economic analysis providing the detail of break-even analysis to identify economically sustainable production volumes.

6.1 Framework for Strategic Alignment

The value of AM technologies and their expected performance are discussed throughout the book to help guide a company on which technology to select as a manufacturing alternative. In fact, selecting a technology should be aligned with a set of factors. The framework in Fig. 6.1 aims to outline the steps that companies should follow to decide whether to adopt AM technology based on their companies' specific context and how to enhance its competitive value through correct selection and implementation. The next chapter provides a discussion to analyze the technology and propose the procedure for selecting a suitable technology from the various AM systems available taking into account the effects of its implementation.

Business strategy is at the core of these considerations, as it should be aligned with the competitive market/industry structure and organizational factors. A corporation then has to determine its competitive priority.

© Springer International Publishing AG 2018
M. Khorram Niaki and F. Nonino, *The Management of Additive Manufacturing*,
Springer Series in Advanced Manufacturing,
https://doi.org/10.1007/978-3-319-56309-1_6

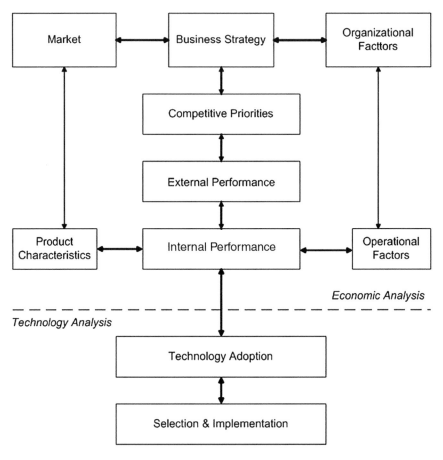

Fig. 6.1 Framework for the strategic alignment of AM technology with business and operations strategy

A business strategy explains the long-term plan of action that a company may pursue to achieve its goals. To do so, a company should link its goals with the capabilities of a technology. The demand or potential market for a product should be linked to the amount of investment in the technology, with less utilization of a technology resulting in reduced profitability of investment. Therefore, companies should firstly identify the competitive environments in which the technology can be most advantageous. As we will see in Sect. 6.2, AM technology can be considered as a highly effective manufacturing method in the following circumstances:

- For developed countries due to its less intense use of resources and corresponding costs.
- For small and medium enterprises because of their flexibility.
- For small to medium production volumes due its compatibility.

- For rapid manufacturing due to a broader range of demands.
- For producing parts made of metal due to their higher value.
- For parts with complex shapes, customization requirements, and creative designs.

Business strategy guides the operations strategy by determining competitive priorities, namely the overall performance objectives able to determine a competitive advantage because it would be evaluated by the market. Therefore, manufacturing extends the business strategy to the operative functions: in other words, it determines how business units deploy their resources to achieve the so-called "order-winning" performance. Decisions taken as part of a company's operations strategy are strategic, because they affect the long-term development of operations resources and processes, and are the basis of a sustainable advantage (Slack and Lewis 2003). Competitive priorities are different for different companies and the most important competitive priorities for each application sector taking into account AM capabilities are listed below:

- *Aerospace*: operations efficiency, weight and material cost savings.
- *Motor vehicles*: operations efficiency, weight and material cost savings.
- *Healthcare*: functions integration.
- *Consumer Products*: customization.
- *Industrial Applications*: operations efficiency, weight and material cost savings.
- *Architecture*: functions integration.
- *Government/Military*: operations efficiency, weight and material cost.

Companies pursue the expected external performances through a set of or a balance set of internal performance. The impacts of AM technology on performance are discussed in detail in Chap. 5, and are categorized in terms of cost, time, quality, and business flexibility. However, these performances are linked to the specific organizational, operational and product characteristics (see Sect. 6.2 for detail).

Analyzing the conditional performance will help companies to decide whether or not adopt AM technology. This analysis is presented in the following paragraphs.

Figure 6.2 illustrates a complete vision of the relationship between expected performance and operational, organizational, and product characteristics, which is an extended framework based on the study of the Khorram Niaki and Nonino (2017b).

However, companies adopting AM have found many applications for the opposite side of these circumstances, and currently these areas are perceived as the most advantageous. In the implementation process, the corporate strategy can be linked to the technology's capabilities and benefits. This method of implementation is called technology-push. Innovation and invention are created in research and development units, and then the capabilities of a technology fabricate the object. With this method, technology may provide opportunities for accessing new markets and/or increasing customers in existing markets. Nevertheless, the newly launched product may or may not satisfy a user need, as happens in the case of many new products. Although the risk of investment during R&D activities can be very high with this method, the acquisition of new markets will certainly increase.

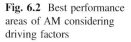

Fig. 6.2 Best performance areas of AM considering driving factors

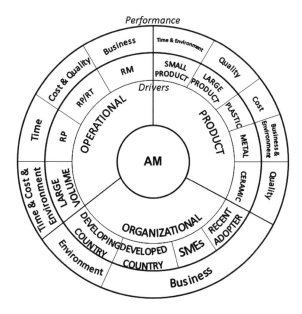

At the core of AM's capabilities is the freedom of design, which enables the production of extraordinary, fully customized and creative products. This may lead to the exceeding of current customer requirements and fulfilling their future needs by offering new functionalities in a product. In addition, developing new creative products requires a high level of investment, while AM technology simplifies this process with a minimum required investment. Consequently, it results in lower concerns about the uncertainty in dealing with product performance and market feedback. Moreover, collaborative design with consumers will further control this risk.

Traditional manufacturing paradigms were based on mass production and the logic of technology-push towards customers, which drove R&D, manufacturing and marketing. The AM adoption process can follow the market-pull strategy in which the technology benefits have to be linked to the requirements of the manufacturing units, and manufacturing requirements are derived from business strategy and customer needs. The schematic of the two approaches (technology-push and market-pull) regarding NPD and innovation processes is illustrated in Fig. 6.3.

According to Baumers et al. (2016), technology-push approaches generally occur in rapid manufacturing, where the technology is used to produce end-usable products, while the use of AM for prototyping follows the demand-pull strategy. Table 6.1 compares the attributes of these two models of implementation. The comparisons are based on several criteria including the uncertainty of dealing with the technology itself, the level of investment required, the required period for R&D activities, the level of uncertainty in dealing with markets, the benefits of reducing the time to market, the possibility of collaborative new product development with customers, the model of market research, and the need for business models to adapt.

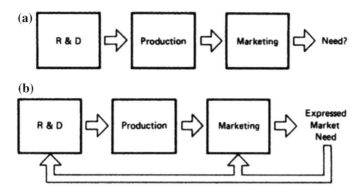

Fig. 6.3 Innovation as technological mutations **a** technology-push; **b** market-pull. *Source* Martin 1994

Table 6.1 Consequences of the transition to AM, technology-push and market-pull strategies

	Rapid prototyping (demand-pull)	Rapid manufacturing (technology-push)
Uncertainty of technology	Low	High
Required R&D funding	Low	High
Duration of R&D programs	Short	Long
Market related uncertainty	Low	High
Time to market	Known	Uncertain/Unknown
User integration into R&D	Easy	Difficult
Mode of market research	Verification	New discovery
Need for user business models to adapt	Minimal	Extensive

Source Baumers et al. 2016

The AM capability in customization also results in a new business model for consumer goods. As seen in Fig. 6.4, moving from a manufacture-centric to consumer-centric method implies a more productive value chain and thus the consumer contributes to value-adding activities. This has not only strategic value but also operational value for manufacturers. For instance, a consumer-centric approach enables a company to pursue co-creation and collaborative design and fabrication of a personalized product, resulting in higher customer satisfaction. It may also offer operational benefits such as less inventory and lower operational costs. The complete characteristics of these two approaches are presented in Table 6.2.

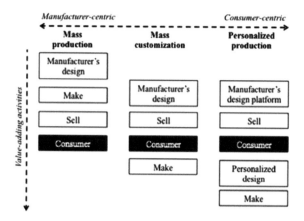

Fig. 6.4 From a manufacturer-centric to a consumer-centric value logic. *Source* Bogers et al. 2016

Table 6.2 Comparison of manufacturer-centric and consumer-centric business models

	Manufacturer-centric	Consumer-centric
Efficiency	• Process transparency	• Low inventory cost
	• Economies of scale	• Print on demand
	• Quality monitoring	• Low operating cost
		• Model reuse
Complementarities	• Portfolio-centric product development	• Indirect linkage to portfolio and product designers
	• Designer creativity	• Multi-partner platforms
Lock-in	• Direct relation to product portfolio	• Support in creation and printing
	• Company-centric community and sharing	• Availability of platforms
		• Community-driven sharing
Novelty	• Freedom for designers	• Co-creation central to design
	• Unique design for special editions	• Personalized designs
	• Co-creation optional	• Localized markets
		• Optional subscription

Source Bogers et al. 2016

6.2   Contingency Factors Driving AM Performance

This section is looking for the generalizability of the impacts, or any explicit links between the impacts and contingency factors. These potential factors are based on three categories, namely organizational factors (country development level, size of the company, and experience in using the technology), operational factors (production volume and objectives of implementing the technology), and product characteristics (product size, and types of raw material used).

6.2.1 Organizational Factors

Country development level

Several studies argued that the influence of Advanced Manufacturing Technology (AMT) on companies in developing countries might vary in comparison with that on companies in developed countries (e.g. Ghani et al. 2002). In addition, research has revealed that the influence of technology on business and organizational structure in a developing country differs from that in a developed country. AM can be considered as a strategic technology for creating value added parts and creating jobs for such developed economies. Generally, developed countries recognized the effectiveness of AM technology on business flexibility more than developing countries do, however, the lower number of respondents from developing countries may result in an incorrect conclusion.

Looking at Fig. 6.5, an interesting observation is that AM resulted in developed countries increasing their competitiveness and innovativeness rather than those in developing countries. This is due to the higher attention paid by developing countries to mass production, and less to technology-oriented production. Mass production requires lower operational costs, with the outcome being low-value products. Thus, implementing AM technology may not be very helpful to their competitiveness. It can improve the competitiveness and flexibility of firms from developed countries by reducing operational costs and increasing innovativeness. Comparing operational costs such as workforce expense in developing countries, we see that it is lower than in most of the developed countries and as a result AM technology can help developed countries to compete due to a reduced demand for labor. Consequently, it is suggested that developed countries should implement AM focusing on high-value products such as those with complex designs and customized and exclusive parts.

Moreover, developing countries recognized a greater effectiveness of AM in environmental sustainability. This may be due to the current status of equipment and technologies being used in developing countries where facilities are probably less environmentally friendly, and thus AM should result in a better experience for them. Likewise, this is the case with energy consumption. Usually, industries in developing countries have a greater energy demand than those advanced systems in developed countries. Consequently, it is concluded that AM will have a greater environmental effect in developing countries, and less of an effect in developed countries. Developing countries also identified a greater effectiveness of AM in accessing new markets and increasing customers. AM was introduced later in developing countries, and as a result it has not yet spread that far. Therefore, in these regions the AM marketplace is less competitive, and new entrants may find many business opportunities.

Enterprise's Size

Studies suggested that small businesses are not scaled-down versions of larger ones and opportunities for large enterprises may not be suitable for small businesses (e.g. Schubert et al. 2007; Federici 2009). Therefore, the benefits of, and barriers to, the adoption of AM by SMEs are likely to be different from those of large companies.

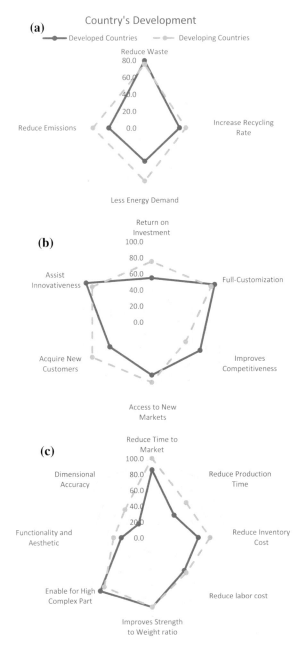

Fig. 6.5 Impact of countries' development level on AM performance: **a** Environmental sustainability performance; **b** Business flexibility; **c** Operational performance

Our survey results demonstrate that despite the availability of some high-tech systems, AM technologies have been considered as viable alternatives in small and medium companies. It also seems that SMEs may benefit more through using the technology, due to their higher flexibility with regard to making changes in their manufacturing and business.

Survey respondents from SMEs recognized a greater effectiveness of AM on business flexibility than those from large companies. Looking at Fig. 6.6c demonstrates that for performance related to time and cost, large companies are in a better situation, but in terms of business flexibility SMEs have more opportunities.

SMEs in comparison with large companies could extend their market and acquire a broader range of new customers if they implement AM. The advent of AM brought opportunities to lots of start-ups and small businesses. As discussed in Sect. 4.1.3, one of these new markets available to small companies is that of 3D printing platforms. Research suggests that the main benefits to SMEs are the local production of highly customized products, facilitations in logistics management, flexibility and a potential decrease in production costs (e.g. Petrovic et al. 2011; Da Silva 2013). Moreover, small companies can focus more on customized products due to their limited customer target groups. Providing more customized products therefore results in the acquisition of new customer from existing markets. Moreover, adopting AM technology has been highly effective in enabling SMEs to increase their competitiveness and SMEs are consequently highly encouraged to adopt AM technologies to help develop their business.

In contrast, the main barriers relate to the cost of AM machines, which can be relatively expensive for SMEs. Another barrier facing SMEs is the need for highly skilled personnel able to master the techniques necessary for dealing with AM such as 3D CAD (Computer-aided Design), solid modeling and process planning. Therefore, in some cases it might be difficult for SMEs to take full advantage of these technologies in the product development process.

Experience in implementing the technology

The maturity of a technology in an organization is an important parameter for evaluation of the profitability of that technology. Studies demonstrated that the structure of most manufacturing firms is evolutionary, thus the fit between technology and structure results in the improved performance of advanced manufacturing technologies. On average, companies, using advanced manufacturing technologies for more than five years have marginally higher performance scores than those of earlier adopters (Small and Yasin 1997).

Our results show the increasing attractiveness of AM adoption in recent years. We considered less than 5 years of experience "Recent" and more than theses years "Former" adopters.

Based on our survey (Fig. 6.7), there are no significant differences in performance relating to the experience of the company. However, a slight difference in quality performance can be seen. Former adopters, which have more experience, saw a slightly higher level of quality from additively manufactured parts, particularly with respect to dimensional accuracy and aesthetics. This is due to the

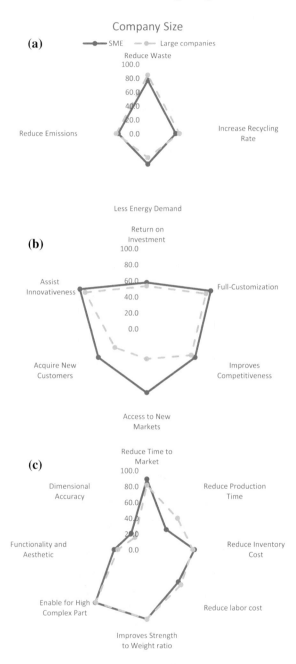

Fig. 6.6 Impact of a company's size on AM performance: **a** Environmental sustainability performance; **b** Business flexibility; **c** Operational performance

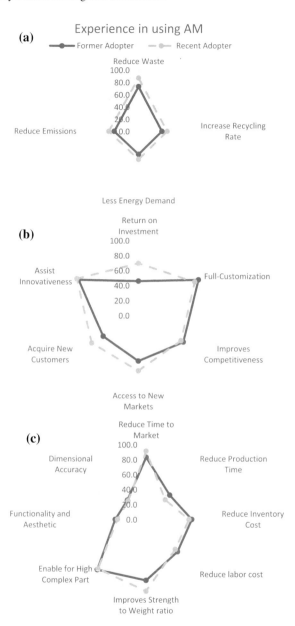

Fig. 6.7 Impact of a companies' experience on AM performance: **a** Environmental sustainability performance; **b** Business flexibility; **c** Operational performance

philosophy of Learning by Doing, which led adopters to learn during iterations, in other words, in post-processing or selection part orientation, which are more skill-based areas. In contrast, recent adopters identified a higher level of effectiveness in other areas apart from quality. This may be due to them benefiting from newer technological advancements or from being in new emerging marketplaces.

6.2.2 Operational Factors

Production Volume

Several studies reveal the profitability of AM technologies for small to medium production volumes (lot size) (e.g., Khorram Niaki and Nonino 2017a) as using AM for large production volumes usually results in expensive parts, mostly because of the slower production speed. For instance, case studies propose that fused deposition modeling (FDM) might be more efficient for low volumes of production. Based on the break even analysis (see Sect. 6.3 for more detail) we found three categories of production volume. These categories are: Small production volume $[\leq 40$ parts], Medium $[\leq 200$ parts], and Large $[>200$ Parts].

As seen in Fig. 6.8, the results of our survey demonstrate the higher efficiency of the technology at larger production volumes (greater than 200 in number). This mostly relates to time and cost, while there are no significant differences in business flexibility. In addition, only about 11% of the survey participants were from companies using AM for large production volumes, and thus it may lead us to an incorrect conclusion.

AM is still seen as an efficient method compared to conventional manufacturing for producing single parts or small volumes. In fact, a production volume of greater than 200 in number is still small in comparison with mass production. The fixed costs (molds, fixtures, etc.) in conventional manufacturing methods result in inefficient methods for single parts or small volume production, but these methods are economical for mass production. Accordingly, when considering the toolless nature of AM, the operational costs of small and medium production lots are expected to be lower than with conventional methods.

Objective of Implementation

Generally, AM is implemented for three main reasons, namely prototyping, tooling and manufacturing. Rapid manufacturing (RM) is defined as the use of AM technology to construct parts that are used directly as finished products or as components. Although the majority of industries still use AM for rapid prototyping (RP) and rapid tooling (RT) to produce prototypes and tools for traditional manufacturing, AM machinery is currently developing its capability in the manufacture of end-usable parts.

As seen in Fig. 6.9b, companies using AM for prototyping or tooling had higher returns on investment in comparison with those using it for the manufacturing of usable final products. This may be due to the lower investment required for prototyping objectives. Rapid prototyping with AM technologies usually requires

Fig. 6.8 Contingency impact of production volume on AM performance:
a Environmental performance; **b** Business flexibility; **c** Operational performance

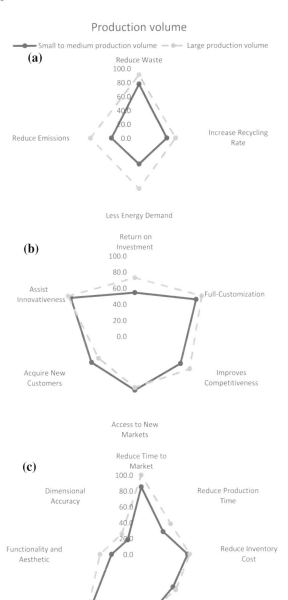

Fig. 6.9 Impact of
implementation objective on
AM performance:
a Environmental
sustainability performance;
b Business flexibility;
c Operational performance

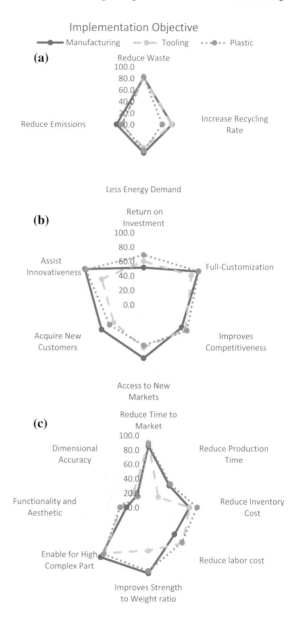

cheaper equipment and raw materials, whereas using the system for manufacturing
requires a more precise, higher production capacity, and the exact raw materials for
the usable parts. Thus, it is clear that with respect to AM, the manufacturing process
needs more investment than prototyping or tooling. Therefore, the results demonstrate a lower return on investment for rapid manufacturing and higher one for rapid
prototyping and rapid tooling.

In addition, those respondents using AM for RP saw a higher product quality level as prototypes require a lower level of quality than final products. Therefore, current AM technologies can fully meet these requirements and further technical advances are needed to move the application toward rapid manufacturing.

In contrast, the survey results show the attractiveness of RM to business. As seen in Fig. 6.9b, those using AM for RM saw a higher effectiveness in business, particularly in the acquisition of new markets and customers. Obviously, RM has a broader range of customers and consumers, while RP only relates to producers. Therefore, RP and RT are not attractive to the market. This illustrates the huge attractive potential of RM if the technological barriers such as production cost and product quality can be overcome.

6.2.3 Product Characteristics

Product Size

The effectiveness and feasibility of AM also depends on the product size. In fact, the primary cost driver of a component is its size and not the production time required as occurs in conventional mass production systems (Achillas et al. 2015). Since there are no explicit measures for this factor, we investigated commercial AM machines in-depth. For the size of products made by AM, the dimensions of the build chamber of the available 3D printers made by two famous 3DP manufacturers (Stratasys and 3D Systems), are taken out. Then, we identified the product size categories as Small [$\leq 250*200*200$ mm], Medium [$\leq 500*400*400$ mm], and Large [>500*400*400 mm], using the mode of the listed dimensions.

As can be seen in Fig. 6.10, differences in product size have an affect only on the performance relating to time and cost, and do not influence business flexibility performance. The results of this survey reveal that respondents using AM for a small product size recognize the greater effectiveness of the technology in reducing production time in comparison with a larger product size.

If the product size becomes larger, then the production time is also longer. However, this longer production time does not mean higher labor costs, since the adding of layers is dependent on the machine and not the operator.

Moreover, results demonstrate that respondents using AM at small product sizes also recognized a greater effectiveness of the technology in reducing energy consumption compared to those with larger product sizes. This is because energy consumption relates to the hours that the machine works. Thus, AM has a greater energy demand whenever the product size becomes larger, while this is not the case in conventional manufacturing. Consequently, those companies producing small-size products are highly encouraged to implement AM in order to reduce both production time and energy consumption. In contrast, the study shows a different trend regarding product quality level with companies implementing AM technology for production of large-size products scoring the effectiveness of AM higher in terms of improving the dimensional accuracy and aesthetics of the final product.

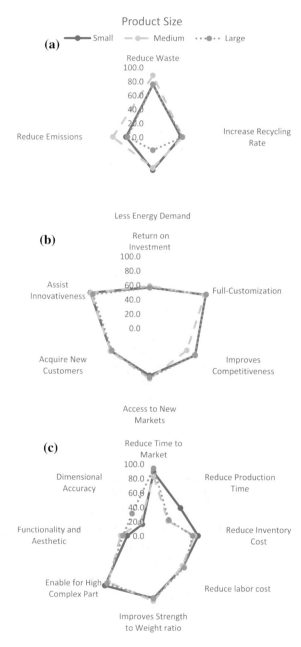

Fig. 6.10 Impact of product size on AM performance: **a** Environmental sustainability performance; **b** Business flexibility; **c** Operational performance

Types of Material

The type of material can influence the performance of AM as the quality and cost of the additively manufactured part depends on the material. Our results demonstrate the widespread application of plastics in AM.

As seen in Fig. 6.11, with regard to time and cost, the best performance of AM occurs in products made of plastic, while the best performance in dealing with the market and customers relates to products made of metal. A case study conducted by Khorram Niaki and Nonino (2017b) is also demonstrating the profitability of investment of metal parts. The best performance of AM regarding the quality of products relates to ceramic materials.

With regard to the return on investment (ROI) indicator, most of the companies using AM for producing parts made of plastic identified a greater effectiveness of the technology for profitability of investment. This is due to the lower initial investment required for the plastic process. The simpler production of plastic using AM also decreases labor costs. In fact, it is even feasible for beginners or less-skilled people to run an AM machine to produce plastic parts. However, most of the companies using AM for producing parts made of metals identified a greater effectiveness of the technology in increasing the recycling rate of materials, with the score for plastics and ceramics being lower than that of metal.

The trend regarding product quality is different from that of cost performance since most of the companies using AM for producing parts made of ceramic recognized a greater effectiveness of the technology in improving product quality in terms of dimensional accuracy and final aesthetics. The lowest quality performance relates to plastic material. The studies explore four interesting relationships relating to quality performance. With respect to parts made of plastics, our findings reveal that the technology needs further advances—both in material properties and 3D printers. The lower quality level often arises from an inappropriate dimensional accuracy or a poor surface finish on parts made of plastic.

Although, there are not any significant differences in business flexibility performance, the processing of metallic materials with AM seems more attractive for developing business. For instance, the best performance of AM in acquiring new market and new customers occurs with metallic materials.

A summary of the findings of this survey are presented in Table 6.3. This table shows the most effective areas, linked to the contingency factors. AM has the greatest effectiveness for business flexibility in developed countries, while it is most beneficial for the environment in developing countries. However, there are no significant differences in other areas. The effectiveness of AM regarding time performance has been mostly seen in large companies, while the best business performance occurs in SMEs.

The application of AM for RP and RT results in advantages in time, costs and quality, although RM is more attractive to business. The effectiveness of AM for business flexibility has been mostly recognized with small to medium production volumes. However, larger production volumes (greater than 200) have better time and cost performance when using AM technologies.

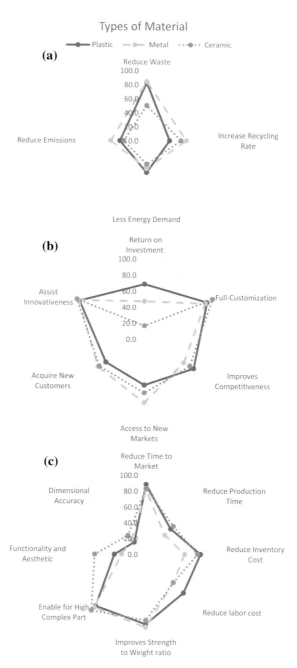

Fig. 6.11 Impact of material on AM performance: **a** Environmental sustainability performance; **b** Business flexibility; **c** Operational performance

Table 6.3 Best performance area of AM considering driving factors

	Drivers	Time performance	Cost performance	Quality performance	Environmental sustainability	Business flexibility
Organization	Country development level	–	–	–	Developing	Developed
	Enterprise size	–	–	–	–	SMEs
	Experience	–	–	–	–	Recent
Operational	Production volume	Large	Large	–	Large	–
	Objective of implementation	RP	RP/RT	RP/RT	–	RM
Product	Product size	Small	–	Large	Small	–
	Types of material	–	Plastic	Ceramic	Metal	Metal

A small product size is perceived to be more suitable for improving time and environmental performance, while a larger product size seems to have better quality. Using metallic materials leads to a better performance in terms of time, environment, and business flexibility, while plastic materials have a better cost performance and ceramic materials have a better quality performance.

However, it is hard to generalize from these findings and comment on the future of the technology since every day we see technological advances in equipment and materials. In addition, every day we also see the successful application of AM in new sectors. Therefore, most of these findings are only relevant to the current status of AM technology.

6.3   Economic Analysis

In order to understand the size of cost-effective production volume, break-even points can be considered as a useful measure. This measure is generally used to determine when a business will be able to cover all its expenses and begin to make a profit. In this case, the break-even points tell us the maximum size of production volumes below which AM technologies represent a more cost-effective manufacturing process compared to conventional manufacturing. However, the break-even point of production is case-specific, depending on technological aspects such as part complexity, material and build volume.

In an early attempt to compare the cost per part of additive and conventional processes, Hopkinson and Dickens (2003) studied SLS (additive) and injection molding (conventional). The curve of injection molding (Fig. 6.12) clearly shows cost-effective production at larger volumes because the initial costs (for instance of the mold) are amortized across the production, whereas the cost per part for the additive process remains constant from a single production to a larger production.

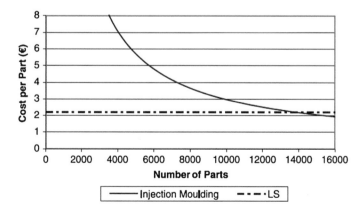

Fig. 6.12 Break-even analysis comparing laser sintering with injection molding. *Source* Hopkinson and Dickens 2003

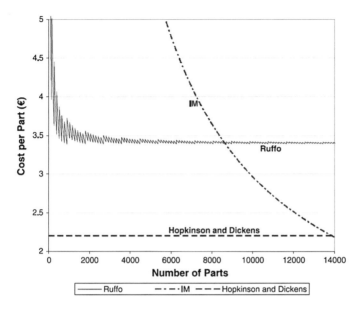

Fig. 6.13 Comparing cost models of laser sintering and injection molding. *Source* Ruffo et al. 2006

However, they studied the RM application of additive processes, which is the production of copies of the same product.

Ruffo et al. (2006) overcame the limitations of the previous study, by considering further components of the cost of production such as labor, materials, machine, and administrative overheads. In contrast to the last study, the new analysis has a curve with a saw-tooth shape to the filling of the machine bed space (Fig. 6.13). It also shows the higher cost per part for single or lower production volumes due to the initial costs of setting up. The break-even point moved from 14,000 parts in the previous study of 2003–9000 parts in the current study, thanks to the newly developed cost estimation model.

Then, Ruffo and Hague (in 2007) looked at the capability of AM technologies in the simultaneous production of different components, which may affect the cost per part at lower production volumes. In other words, as seen in Fig. 6.14, the study compared the production of copies of the same parts and parallel printing of mixed components, both with an SLS system. The results demonstrate the efficient operational costs when simultaneously printing mixed components. The optimized mixing of the components in the building platform also results in further cost savings.

The previously mentioned experiments considered objects that were additively manufactured with plastic materials. So, for parts made of plastic, the results demonstrate that the competitiveness of AM in comparison with conventional manufacturing is well established. However, when considering the design optimization offered by AM, more cost effective operation can be expected as previous studies did not consider the possible optimization, and the design file was that used in conventional manufacturing. The potential design optimization comes from the

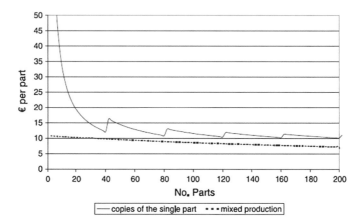

Fig. 6.14 Comparing cost models of copies of single products and mixed components. *Source* Ruffo and Hague 2007

freedom of design to produce lightweight parts, complex structures, integrated parts, hierarchical structures, and so on.

As discussed in Chap. 5, designers and engineers should consider the following items if they are to achieve an optimum design with AM (Gibson et al. 2010):

- The possibility of using undercuts, flexible wall thickness and deeper channels
- The possibility of fabricating parts with unlimited complexity such as twisted shapes, blind holes, and screw forms
- The possibility of combining several components into an integrated assembly.

With these aspects in mind, in an effort to evaluate the efficiency of AM production of metal parts, Atzeni and Salmi (2012) conducted an analysis involving the comparison of the SLS of an additive process with high-pressure die-casting in a conventional process, to look at possible design optimization. They employed an EOSINT M270 machine (from EOS GmbH) and direct metal laser sintering (DMLS). The sample part, an aeronautic component, (Fig. 6.15a) is the main landing gear of the Italian P180 Avant II aircraft, built by Piaggio Aero Industries S.p.A. The redesigned part (Fig. 6.15b) benefits from a reduction in the number of components that need assembling and a reduction of as much as possible in material usage through the use of undercuts and hollow structures, and a design for optimum functionality.

The cost components usually include materials, processing and tooling. Processing costs consist of machine amortization, design and testing, labor, and both pre- and post-processing. Tooling cost is only considered for traditional manufacturing, for instance with high-pressure die-casting the cost of an object consists of four main components—the cost of materials, the cost of the mold, the cost of the process and post-processing per part. Likewise, the cost of an SLS-printed part also contains four components: pre-processing, material, processing, and post-processing. The details of the cost per part in both high-pressure die-casting and SLS processes are presented respectively in Tables 6.4 and 6.5.

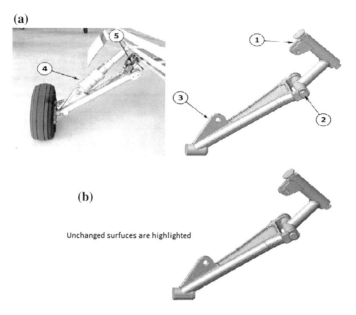

Fig. 6.15 Main landing gear of the Italian aircraft P180 Avant II (**a**); and the redesigned part (**b**). *Source* Atzeni and Salmi 2012

Furthermore, for the former, an assembly operation is required and the cost of the assembly is calculated at the end, based on the production volume (N).

Based on the results obtained and the cost model, Fig. 6.16 illustrates the break-even curves when comparing an SLS additive process with a conventional high-pressure die-cast method for a part made of aluminum alloy. Specifically, SLS can be considered as an efficient manufacturing method when the production volume is lower than forty-two assembly components. However, the observed justifications were more than just the economic aspects. For instance, the total process in the SLS method took only 2.5 days, while pre-processing (such as mold production) took several weeks in the conventional process, contributing in a reduction of time to market.

Table 6.6 shows the results of some other case studies that attempted to find the maximum production volume at which AM can be cost-effective in comparison to conventional processes. Five case studies are presented in the table. For instance, in terms of metallic materials, Lindemann et al. (2012) compared selective laser melting with a traditional subtractive milling process. AM can be considered as a cost-effective manufacturing method for this part when the production volume is lower than 190 in number. Broadly speaking, the maximum cost-effective production volume when using polymers for AM is higher than that for metal parts.

In these studies, the production time was measured experimentally in laboratories. Nevertheless, the ability to estimate the cost of production early is an indispensable attribute and one of the main issues in the preliminary evaluation of the technology. Several studies have proposed models for estimating the build time and cost of products manufactured using AM technology based on artificial

Table 6.4 Costs for high-pressure die-cast parts

Components	Unit	Amount
Production volume	*(pcs)*	*N*
Material cost per kg	(EUR/kg)	16.00
Part weight (kg)	(kg)	0.162
Material cost per part	*(EUR)*	*2.59*
Standard components' cost	(EUR)	1900
Mold cavities and slides cost	(EUR)	15,400
Ancillary cost	(EUR)	3700
Mold cost per part	*(EUR)*	*21,000/N*
Machine cost per hour	(EUR/h)	260.00
Cycle time	(h)	0.001
Labor cost per hour processing	(EUR/h)	35.00
Percentage of operator time	(%)	10%
Processing cost per part	*(EUR)*	*0.26*
Heat treatment cost per part	(EUR)	1.42
Machining operations cost	(EUR)	13.98
Labor cost per hour post-processing	(EUR/h)	25.00
Operator time	(h)	0.100
Post-processing cost per part	*(EUR)*	*17.90*
Total cost per part	*(EUR)*	*20.75 + 21,000/N*
Linkages parts' costs	(EUR)	0.50
Assembly cost	(EUR)	0.04
Total cost per assembly	(EUR)	21.29 + 21,000/N

Source Atzeni and Salmi 2012
Italics the cost per part

intelligence (Di Angelo and Di Stefano 2010, 2011; Munguía et al. 2009). Their proposed parametric approach analyzes the geometrical features which typically affect the build time of AM.

Obviously, the production cost of using AM technology is linked to the build time, along with other drivers like materials. The build time depends on the characteristics of the additive systems, object geometries and complexity, and build orientation. Build time is usually the summation of the time used for printing layers and the delay between layers forming. Layer printing time is the time that the tool takes to create the layer contours and the tool path loops. The tool path loops depend on the tool path length and the number of tool repositionings. The delay time between forming layers is employed for the solidification and cooling of the deposited layers.

The build time of additive processes usually consists of seven main components as follows (Di Angelo and Di Stefano 2011):

- Scanning time of the material contour
- Time for hatching material

Table 6.5 Costs for part made by SLS

Components	Unit	Amount
Numbers of parts produced per job	*(−)*	4
Material cost per kg	(EUR/kg)	145.00
Part volume	(mm³)	0.060
Density of the sintered material	(g/mm³)	2.68
Mass of material per part	(kg)	0.178
Material cost per part	*(EUR)*	*25.81*
Machine operator cost per hour	(EUR/h)	20.00
Set-up time per build	(h)	1.2
Pre-processing cost per part	*(EUR)*	*8.00*
Machine cost per hour	(EUR/h)	35.00
Build time	(h)	54
Machine cost per build	(EUR)	1890.00
Processing cost per part	*(EUR)*	*472.50*
Machine operator cost per hour	(EUR/h)	20.00
Post-processing time per build	(h)	13.98
Heat treatment cost per build	(EUR)	20.00
Post-processing cost per part	*(EUR)*	*20.00*
Total cost per assembly	(EUR)	526.31

Source Atzeni and Salmi 2012
Italics the cost per part

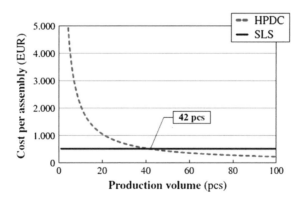

Fig. 6.16 Break-even analysis comparing conventional HPDC process with SLS. *Source* Atzeni and Salmi 2012

- Scanning time of the supports' contours
- Time for hatching supports
- Repositioning time of deposition tool
- Repositioning time of the supports' deposition tool
- Delay time between deposition of subsequent layers.

Table 6.6 Break-even points of AM technologies compared to conventional manufacturing

Break-even point (in pieces)	Printed material	Process comparison	Source
279–5800	Polymer	SLA compared to injection molding	Hopkinson et al. (2006)
7500	Polymer	FDM compared to injection molding	Hopkinson et al. (2006)
14,000	Polymer	SLS compared to injection molding	Hopkinson et al. (2006)
42	Aluminum	SLS compared to high-pressure die-casting	Atzeni and Salmi (2012)
190	Steel	SLM compared to milling	Lindemann et al. (2012)

Table 6.7 Sub-processes involved in typical commercial AM technologies

Systems	Sub-Process			
	Material		Support	
	Border scanning	Area scanning	Border scanning	Area scanning
SLA	YES	YES	YES	YES
SLS	YES	YES	NO	NO
SLM	YES	YES	NO	NO
FDM	YES	YES	YES	YES
LOM	YES	NO	NO	YES
3DP	NO	YES	NO	YES
EBM	YES	YES	NO	NO

Adapted from Di Angelo and Di Stefano 2011

However, as discussed in Chap. 1 the processes of AM technologies are different, resulting in different processing time components. For instance, as seen in Table 6.7 the sub-process involved in typical commercial AM technologies, which are the driving factors of processing time.

In parametric build time estimation, driving factors play an important role in the process. These factors can be identified through the STL standard definition of the geometric model. Eight parameters (V_{mat}/L_{mat}, b_{z-mat}/L_{mat}, p_{mat}, n_{r-mat}, V_{sup}/L_{sup}, b_{z-sup}/L_{sup}, p_{sup}, n_{r-sup}) consider the dimensional features of the object and the dimensional technological parameters.

where:

V_{mat}	Volume of material to be printed; V_{sup} (for support)
L_{mat}	Layer thickness; L_{sup} (for support)
b_{z-mat}	Object's height; b_{z-sup} (for support)
p_{mat}	Total length of the layers' contour to be deposited; p_{sup} (for support)
n_{r-mat}	The number of repositioning movements; n_{r-sup} (for support).

Using the above-mentioned estimation model, companies can estimate the build time and can then estimate the production cost of a part with AM. Therefore, they can see whether or not the AM technologies are cost-effective when compared with the existing manufacturing methods in the initial stages of implementing the technology.

Exhibit 14—Economical Analysis—Return On Investment
Return on investment (ROI) takes into account the gain or loss generated on an investment relative to the amount of money invested. It allows one to compare the efficiency of different investments. Therefore, one parameter for comparing AM with conventional manufacturing is to know the profitability of any investment. Studies and practitioners confirm the benefits of AM technology in terms of production and business opportunities, because since AM enables toolless production and needs a smaller production line, it can reduce operational costs. AM may offer a much better investment opportunity in comparison with conventional manufacturing, particularly for high-value parts with low-volume production.

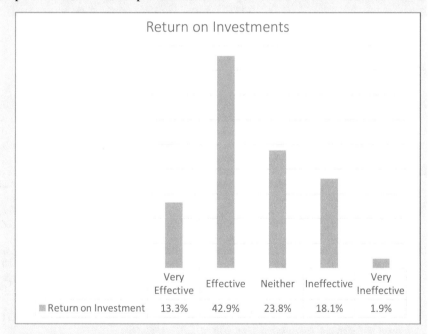

	Very Effective	Effective	Neither	Ineffective	Very Ineffective
■ Return on Investment	13.3%	42.9%	23.8%	18.1%	1.9%

More than 43% of the respondents believe that AM is not good for profitability of investment. This may be due to the higher initial investment in equipment in some circumstances such as the production of parts made of metal or those with a need for greater precision. AM machineries have to fully consider this performance indicator so as to reduce the amount of investment required by as much as possible. Notably, AM adopters can also think of creating new business with those assets and/or should consider how to take full advantage of the technology's potential.

Seven Key Facts

- AM can drive competitiveness for developed countries because it needs fewer resources.
- AM is more advantageous for small and medium enterprises because of their flexibility.
- AM is an efficient manufacturing method for small to medium production volumes.
- AM is an effective manufacturing method for parts with complex shapes, customization requirements, and creative designs.
- The operational performance using AM for prototyping is more efficient than manufacturing end-usable products.
- The operational performance when using AM for small-size products is more efficient than large ones.
- Approximately 57% of AM adopters believe that AM is better than conventional manufacturing in terms of profitability of investment.

References

Achillas, C., Aidonis, D., Iakovou, E., Thymianidis, M., & Tzetzis, D. (2015). A methodological framework for the inclusion of modern additive manufacturing into the production portfolio of a focused factory. *Journal of Manufacturing Systems, 37,* 328–339.

Atzeni, E., & Salmi, A. (2012). Economics of additive manufacturing for end-usable metal parts. *The International Journal of Advanced Manufacturing Technology, 62*(9–12), 1147–1155.

Baumers, M., Dickens, P., Tuck, C., & Hague, R. (2016). The cost of additive manufacturing: Machine productivity, economies of scale and technology-push. *Technological forecasting and social change, 102,* 193–201.

Da Silva, J. V. (2013, October). 3D technologies and the new digital ecosystem: A Brazilian experience. In *Proceedings of the Fifth International Conference on Management of Emergent Digital EcoSystems* (pp. 278–284). ACM.

Di Angelo, L., & Di Stefano, P. (2010). Parametric cost analysis for web-based e-commerce of layer manufactured objects. *International Journal of Production Research, 48*(7), 2127–2140.

Di Angelo, L., & Di Stefano, P. (2011). A neural network-based build time estimator for layer manufactured objects. *The International Journal of Advanced Manufacturing Technology, 57*(1–4), 215–224.

Federici, T. (2009). Factors influencing ERP outcomes in SMEs: A post-introduction assessment. *Journal of Enterprise Information Management, 22*(1/2), 81–98.

Ghani, K. A., Jayabalan, V., & Sugumar, M. (2002). Impact of advanced manufacturing technology on organizational structure. *The Journal of High Technology Management Research, 13*(2), 157–175.

Gibson, I., Rosen, D. W., & Stucker, B. (2010). *Additive manufacturing technologies.* New York: Springer.

Hopkinson, N., & Dicknes, P. (2003). Analysis of rapid manufacturing—Using layer manufacturing processes for production. *Proceedings of the Institution of Mechanical Engineers, Part C: Journal of Mechanical Engineering Science, 217*(1), 31–39.

Hopkinson, N., Hague, R., & Dickens, P. (Eds.). (2006). *Rapid manufacturing: An industrial revolution for the digital age.* Wiley.

Khorram Niaki, M., & Nonino, F. (2017a). Additive manufacturing management: A review and future research agenda. *International Journal of Production Research, 55*(5), 1419–1439.

Khorram Niaki, M., & Nonino, F. (2017b). Impact of additive manufacturing on business competitiveness: A multiple case study. *Journal of Manufacturing Technology Management, 28*(1), 56–74.

Lindemann, C., Jahnke, U., Moi, M., & Koch, R. (2012, August). Analyzing product lifecycle costs for a better understanding of cost drivers in additive manufacturing. In *23rd Annual International Solid Freeform Fabrication Symposium—An Additive Manufacturing Conference.* Austin Texas USA 6th–8th August.

Munguía, J., Ciurana, J., & Riba, C. (2009). Neural-network-based model for build-time estimation in selective laser sintering. *Proceedings of the Institution of Mechanical Engineers, Part B: Journal of Engineering Manufacture, 223*(8), 995–1003.

Petrovic, V., Vicente Haro Gonzalez, J., Jorda Ferrando, O., Delgado Gordillo, J., Ramon Blasco Puchades, J.,& Portoles Grinan, L. (2011). Additive layered manufacturing: Sectors of industrial application shown through case studies. *International Journal of Production Research, 49*(4), 1061–1079.

Ruffo, M., & Hague, R. (2007). Cost estimation for rapid manufacturing simultaneous production of mixed components using laser sintering. *Proceedings of the Institution of Mechanical Engineers, Part B: Journal of Engineering Manufacture, 221*(11), 1585–1591.

Ruffo, M., Tuck, C., & Hague, R. (2006). Cost estimation for rapid manufacturing-laser sintering production for low to medium volumes. *Proceedings of the Institution of Mechanical Engineers, Part B: Journal of Engineering Manufacture, 220*(9), 1417–1427.

Schubert, P., Fisher, J., & Leimstoll, U. (2007). ICT and innovation in small companies. In *ECIS* (pp. 1226–1239).

Slack, N. & Lewis, M. (2003), *Operations Strategy*, Prentice Hall.

Small, M. H., & Yasin, M. M. (1997). Advanced manufacturing technology: Implementation policy and performance. *Journal of Operations Management, 15*(4), 349–370.

Chapter 7
Selection and Implementation of Additive Manufacturing

Potential AM adopters can recognize the suitability of AM technology to their circumstance through the analysis provided in pervious chapter. This chapter discusses the technology selection and implementation processes. It begins with an analysis of the technology and then attempts to depict the future of AM technology. It describes the future possible beneficiaries of this technology and introduces the process for AM technology selection. It presents appropriate selection criteria and proposes a process for selecting a system from among the different manufacturing technologies available as well as selecting from different AM technologies. The chapter goes on to describe the implementation process for AM technologies in an organization. A detailed discussion on the changes required in an organization, its operations and supply chains is provided. As for the operational changes, the chapter analyzes the design process, production planning and control, quality control, and the role of the Internet of Things (IoT) in implementation.

7.1 Technology Analysis

As shown in Fig. 6.1, a careful technology analysis should be done after economic analysis. The current technological barriers to the widespread uptake of the technology were outlined in Sect. 4.4. These factors mostly involve the range of available materials, cost of machines, lower finish quality, production time, etc. Therefore, companies should be aware of these issues before adoption.

These factors are probably due to the relative immaturity of the technology. For instance, production time performance is an important factor for an alternative manufacturing method. To choose an AM technology as an alternative to conventional processes, it also needs to be competitive in terms of processing time. However, although it is difficult to predict when AM will become competitive in terms of production time, we can look its evolution to date. Holmström et al. (2016) provided an example of this evolution. Figure 7.1 illustrates the approximation of

© Springer International Publishing AG 2018

M. Khorram Niaki and F. Nonino, *The Management of Additive Manufacturing*,
Springer Series in Advanced Manufacturing,
https://doi.org/10.1007/978-3-319-56309-1_7

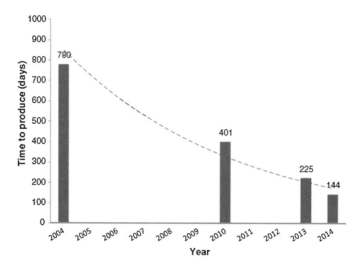

Fig. 7.1 Estimated days to 3D print the steel parts of a passenger car on a single selective laser melting 3D printer. *Source* Holmström et al. 2016

the days required for additively manufacturing the components of a car. The results come from the best available commercial DMLS machines used in four periods (2004, 2010, 2013, and 2014), and producing the same amount (135,000 cm^3) of parts made of steel materials. For instance, 780 days would have been required in 2004, while in 2014 the production times had reduced to 144 days, resulting in a 450% improvement. This is due to the increased rate of material deposition per hour of the current machines. Consequently, we can estimate the same trend in future.

Furthermore, we have seen the efforts made to develop and supply more affordable machines, particularly for micro and small enterprises. Therefore, we can expect an optimistic future for the technology. To this end, Jiang et al. (2017) attempted to predict the future of AM technology through the opinions of sixty-five industry experts. They applied the Delphi method, which is a multi-stage fore-casting method used to identify technical developments and trends, through which a consensus among a set of experts on future developments is identified. Table 7.1 presents a set of eighteen projections that might occur in/or by 2030. These pro-jections are categorized as production and supply chain, business model and competition, consumer and market trends, and intellectual property and policy.

The results of this survey demonstrated the most probable scenario for 2030. Figure 7.2 illustrates the most probable scenarios, categorized by all of the elements of a value chain. Projection 9 (product development moves to an iterative and agile processes) is the most probable scenario for product development in 2030. Projections 5 (in situ production of spare parts) and 14 (larger share of products with multi-materials and/or contain embedded electronics) are the most predictable scenarios in terms of product attributes. In terms of channels of distribution, the most likely scenarios will be the use of online 3D printing platforms by consumers to buy designed files or final 3D printed products (projections 12 and 17). The most

Table 7.1 Delphi projections for 2030. Adapted from Jiang et al. 2017

Areas	Pro No.	Projections in/by 2030
Production, supply chain, and localization	1	50% of overall industrial AM will be in-house production capacity
	2	A significant amount of SMEs will share AM production resources (industry-specific) to achieve higher machine utilization, learning effects, quality assessments, etc.
	3	Across all industries, local AM production near customers will increase significantly, resulting in a de-globalization of supply chains
	4	Distribution of final products will move significantly (N25%) to selling digital (design) files instead of selling the physical product
	5	Fewer critical spare parts will be locally produced by AM technology, whereas critical spare parts will be made at specialist hubs
	6	AM substantially reduces the carbon footprint of manufacturing and transportation
Business models and competition	7	AM technology will shift the sources of competitive advantage from manufacturing and supply chain capabilities towards access to customer and designer networks
	8	Firms' business models will not be immensely influenced by AM, as it is just another manufacturing technology, requiring novel knowledge and skills
	9	Conventional measures of "time to market," "product lifecycle," and "ramp-up" will be diminished, since the digital products will be subjected to frequent design iterations and constant modifications
	10	Germany will be among the top five global players in developing industrial AM technology and machinery due to existing machine producers, research institutions, and a large number of end users
Consumer and market trends	11	The market share of additively fabricated parts versus conventionally produced articles will be significant (N10%) across all industries
	12	A significant number of consumers will utilize online 3D printing platforms to purchase product designs or for production
	13	The majority of individual consumers (in industrial countries) will have desktop 3D printers at home
	14	A significant amount of AM fabricated products will consist of multi-materials and/or contain embedded electronics, enabling a broad range of applications
	15	AM fabricated human organs will be a viable and largely utilized substitute for donor organs

(continued)

Table 7.1 (continued)

Areas	Pro No.	Projections in/by 2030
Intellectual property and policy	16	AM communities have to significantly use the novel forms of intellectual property like Creative Commons, or open source
	17	An important regulatory measure will be the regulation of AM file-sharing platforms
	18	The problem of liability due to unclear intellectual property rights will have led to a much lower utilization of AM than is technically possible

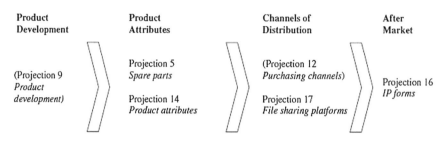

Fig. 7.2 Most probable scenario for additive manufacturing in 2030. *Source* Jiang et al. 2017

likely issues in the market will be defending intellectual property, therefore AM communities have to significantly develop novel forms of intellectual property.

As seen in Fig. 7.2, there are six most likely projections as discussed above, with twelve others that are less likely to happen by 2030. Looking at two of the less probable projections (projection 8: the impact of AM on firms' business models; and projection 12: consumer distribution channels), there will be four possible scenarios in the future (see Fig. 7.3).

- *Enhancing of existing business models; and consumers will shift to the use of online 3DP platforms*
  In this situation, called market explorer, companies should extend their digital file selling. It will also create a new exports paradigm, namely the sale of digital files instead of offering physical objects.
- *Creating of new business models; and consumers will purchase though conventional sales channels*
  In this scenario, called content provider, companies become digital file suppliers only, and the manufacturers will also be the consumers. This will dramatically change the business model so that many factors (for instance, intellectual property, asset allocation, and branding issues, etc.) should be fully taken into account.

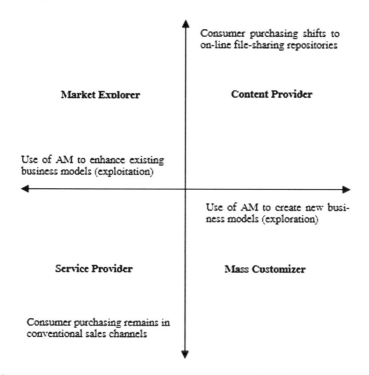

Fig. 7.3 Four controversial extreme scenarios showing how AM affects consumer purchasing models. Adapted from Jiang et al. 2017

In this circumstance, the AM adopter becomes the service provider. Companies will produce small production volumes of product—as is the current status of the AM industry. They will focus on specific products benefiting from AM technology capabilities.

- *Creating of new business models; and consumers will purchase though conventional sales channels*
  This scenario will turn the AM adopter to a mass customizer as they implement AM technology to fulfill the requirement of each individual customer. In addition, AM will remain as a manufacturing method for in-house production, enabling consumers to produce highly varied parts at low complexity and cost. Looking at these probable future scenarios helps AM adopters to be prepared for the future of the technology so that the major changes of the future don't result in them losing the benefits of their existing business models.

7.2 Selecting AM Technology

Different manufacturing methods—both conventional and additive—have their own application spectrum (see Fig. 7.4). AM processes are particularly compatible with concept models of low volume production end-usable products (Khorram Niaki and Nonino 2017a), while conventional manufacturing (in other words die casting and injection molding) are the most suitable manufacturing methods for mass production. However, each different AM technology has its own spectrum, for instance jetting systems (namely multi-jet modeling and binder jetting) are perceived as the most suitable for concept modeling rather than for prototyping or low volume production, while direct metal laser sintering is suitable for prototyping as well as the production of end-usable products.

Technology selection is also influenced by the prototypes and tools categories. There are five prototype categories, namely conceptual model, technical prototype, sand casting, investment casting, and plastic molding (Armillotta 2008).

Conceptual models are the prototypes required before product documentation. These models are commonly used for visual checking, ergonomic testing, and initial sharing with manufacturers or customers. Therefore, there is no need for accurate dimensions or a sophisticated surface finish. In contrast, technical prototypes are used for fit and functional testing and should have relatively accurate geometry and actual material properties. The rest of the prototype models, used in pilot production runs, are called multiple prototypes. These models are used for the final evaluation of a product (cast or molded), which is an ideal preparation for

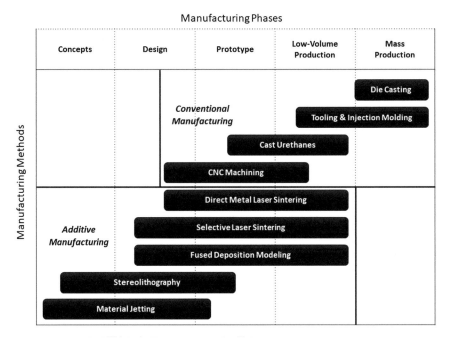

Fig. 7.4 Manufacturing technologies and the application spectrum. Adapted from Cotteleer and Joyce 2014

Table 7.2 Technology selection for prototype categories. Adapted from Armillotta 2008

Alternatives	Prototype categories				
	Conceptual model	Technical prototype	Sand casting	Investment casting	Plastic molding
Stereolithography (SLA)	✓	✓	✓	✓	
Selective Laser Sintering (SLS)	✓	✓	✓	✓	
Fused Deposition Modeling (FDM)	✓	✓	✓	✓	
Lamination Object Manufacturing (LOM)	✓	✓	✓		
InkJet Printing-wax (IJP)	✓			✓	
Multi Jet Modeling - thermoplastic (MJM)	✓				
PolyJet-photopolymer (PJ)	✓	✓	✓		
Direct Metal Laser Sintering (DMLS)	✓	✓			
Room Temperature Vulcanizing (RTV) tooling		✓		✓	✓
Epoxy tooling					✓
SLA tooling				✓	✓
Sintered tooling				✓	✓
CNC Al tooling				✓	✓

rapid tooling options. The model can be applied as a pattern for various manufacturing processes such as sand casting, investment casting, and injection molding. Therefore, these tools should also have the same requirements as the final products such as surface finish and dimensional accuracy. Table 7.2 reports the suitable alternatives for each of these categories based on their specific requirements. Alternatives include a set of conventional manufacturing as well as various AM technologies.

Several academic studies have looked at the use of operational research techniques for decision making in AM management. These studies considered the problem of optimal AM system selection, allowing for the specific process and given product characteristics using different decision-making approaches (e.g., Kengpol and O'Brien 2001; Rao and Padmanabhan 2007; and Lan 2009). Due to the rapid growth of AM technologies, the procedure for selecting the most appropriate technique from a number of AM technologies is becoming increasingly important.

Different AM technologies were explained in Chap. 1. It was remarked that each one has its strengths and weaknesses in terms of both technological and economic aspects, and each one is therefore well suited to a specific product type or production area. The objective of the selection process is a quantitative comparison between the AM techniques available for a given application, or in some studies among AM and other conventional manufacturing techniques. Nevertheless, most users depend on information from experts and service suppliers, rather than on a constructive and formal evaluation procedure for the suitability of an AM technology for a prospective project.

7.2.1   Selection Criteria

A typical selector system contains several options (different AM techniques or commercial brands of 3D printers) and various selection criteria. Thus, the input data may include the type of product, quantity of the part to be manufactured, lead time (which can vary from a few hours to several weeks), and some geometric properties such as size, layer thickness, and required accuracy. Byun and Lee (2005) identified six attributes for the selection and evaluation process, namely accuracy, surface roughness, tensile strength, elongation, cost of the part and build time. The build time includes the pre-processing time, building time and post-processing time. The part cost includes both material and labor costs.

As a further extension, Armillotta (2008) developed the decision criteria, including all relevant and independent requirements of AM techniques for prototyping and tooling processes. Eleven attributes were presented, concerning part properties, production features, and operation costs. These attributes are as follows:

- Fitting to office environment (considering the size, cleanliness, and environmental emissions)
- High build speed (ratio of cm^3 per hour)
- Less need for setup (for example in construction of special tooling)
- Less need for post-processing
- Availability of either functional or high-strength raw materials for the system
- Good dimensional accuracy
- Good surface finish
- Economical processing for larger product sizes and production volumes
- Lower cost of raw materials
- Lower cost of operating the system
- Lower cost of setup (e.g. special tooling).

Figure 7.5 clearly illustrates the hierarchical nature of the optimal selection process with various AM systems. The decision criteria are classified into five categories, namely technology (the parameters dealing with capability of the specific technology), geometry (the geometrical flexibility offered by the AM system), performance (the parameters relating to the mechanical properties of the

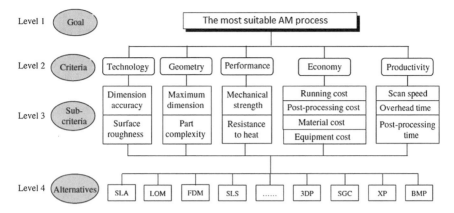

Fig. 7.5 Hierarchical AM technology selection and criteria. *Source* Lan et al. (2005)

fabricated parts), economy (including the total operational costs of using a specific AM system), and productivity (including those parameters dealing with manufacturing time).

7.2.2 Selection Process

The selection process for prototyping or manufacturing, considering the given application's characteristics, has always been an essential decision to be made for users of both conventional manufacturing and AM. Thus, several methodologies, decision-support techniques and software tools to assist decision makers in their choice of the most suitable process, were developed. As with any other optimization process, these systems use such optimization techniques as analytic hierarchy process (AHP) (e.g., Kengpol and O'Brien 2001), expert systems (e.g., Masood and Soo 2002), the fuzzy approach (e.g., Khrais et al. 2011), or graph theory and the matrix approach (Rao and Padmanabhan 2007). However, although this section provides an overview of these academic efforts, it does not go into the details of each system and its optimization process.

Hornberger developed the RP process selector at Santa Clara University in 1993, perhaps the first attempt to develop a computer-based program to select a proper RP system. Campbell and Bernie (1996) then established a database of the RP system capabilities to help RP users make the most efficient use of RP. The outline is founded on the theory that "once the capacity of an RP system has been specified for individual features its capability for any component containing these characteristics can be predicted".

Then, in 2002, Helsinki University of Technology developed a web-based selector. The user of this selection process chooses their requirements through the decision criteria. Then, the system feeds back a list of the three best techniques to fulfil their requirements. However, this program is not perfect—not even close to perfect—but is a friendly pointer in the right direction. Masood and Soo (2002) introduced a rule-based expert system for the selection of an RP system from all the commercial RP systems manufactured in the USA, Japan, Germany, and Israel. Chung et al. (2003) proposed an RP process selector for educational and manufacturing purposes from among a wide range of RP systems commercially available from RP manufacturers worldwide. The search routines give access to a large database built on entity relationship techniques to enable a quick retrieval of information.

Furthermore, a two-stage decision-support system was developed in 2005, in which the first stage utilizes the expert system environment and generates feasible alternatives, and then in the second stage the fuzzy synthetic evaluation approach is used to rank the options. The system provides a platform mounted on the server, accessible to users. This is composed of four modules: (1) a database to store RP processes with their specifications; (2) a knowledge-based expert system for selecting feasible alternatives; (3) a fuzzy model to choose the most suitable RP process; and (4) user and expert interfaces to interact with the system (Lan et al., 2005). Four modules work together to complete the decision-making task. The overall architecture of this web-based arrangement is shown in Fig. 7.6.

Khrais et al. (2011) also presented a fuzzy logic approach to selecting the best RP technique. Evaluation criteria were set and experts were consulted to assign

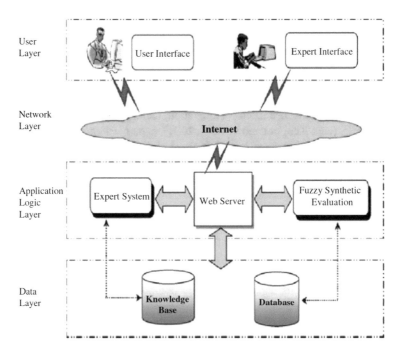

Fig. 7.6 The architecture of a web-based RP selector system. *Source* Lan et al. 2005

importance weights to them. The approach classifies the elements which affect the selection process into static and dynamic factors. Fuzzy if-then rules and fuzzy sets are utilized to translate the appropriateness of each RP technique to each evaluation criterion of the task. The "Max–Min set" method is used to obtain the efficiencies, which are blended with their importance weights.

To understand the procedure and outcomes of the selector systems, an example of the research is detailed below. Roberson et al. (2013) evaluated five home-use desktop AM systems. This evaluation considered important parameters such as build time, material usage, support and waste material, dimensional accuracy and surface roughness. The detailed information of the systems is presented in Table 7.3 along with cost and technical information. Using the results obtained in the laboratory for the parts tested in production volume quantities of one and five, the comparative results in terms of the parameters mentioned are presented.

Comparing build times, 3D Touch was the most time-consuming AM system, taking about 329 min for one part and 1373 min for five parts. The Replicator system had the fastest build time (220 min) when building one part. In terms of material usage, the comparative results demonstrate that the extrusion-based methods consume less material compared to the SD300 Pro (sheet lamination process). The LOM system consumed 1168.9 g of materials, while the 3D Touch (extrusion-based) used only 35.5 g, 97% less than the SD300 Pro. Within the extrusion-based systems, the Replicator system had a slightly higher material usage (42.5 g per part). However, in case of the simultaneous printing of five parts, the V-Flash (vat photopolymerization process) used same amount of materials as the extrusion-based systems. In the analysis of material waste, the SD300 Pro had

Table 7.3 Information relating to desktop 3D printers, used in the study of Roberson et al. (2013)

Developer	Model	Build size (mm)	Build envelop size (cm³)	Cost (USD)	Available layer thicknesses (mm)	Technology
Dimension	uPrint plus	203*203*135	5563.2	20,900	0.254	Material extrusion (FDM)
					0.330	
Solido	SD300 Pro	160*210*135	4536	4375	0.168	Sheet lamination
Bits from bytes	3D touch	275*275*210	15,881.3	3930	0.102	Material extrusion (FDM)
					0.125	
					0.250	
					0.500	
3D systems	V-Flash FTI 230	228*171*203	7914.6	9900	0.102	Vat photo polymerization
MakerBot	Replicator	225*145*150	4839.8	2072	0.200	Material extrusion
					0.300	

the highest amount (1120.4 g when printing a single part). Within the extrusion-based methods, 3D Touch had the highest waste.

Dimensional accuracy was evaluated based on the comparison between original CAD drawings and measurements of the actual printed parts. Parts made with the V-Flash were mostly undersized and were oversized with the Replicator system. Measurements of height demonstrate that parts made with the V-Flash, Replicator and 3D Touch, were greater than what was indicated by the CAD drawings. Considering measurements of symmetry, all units produced features close together compared to the CAD drawings, with all being less than 5% undersized.

Surface roughness was also evaluated (with a Mitutoyo model SJ-201P surface roughness tester). The SD300 Pro and the V-Flash fabricated the smoothest surfaces while in contrast the Replicator and 3D Touch produced the roughest parts. Within the extrusion-based systems, the uPrint had the best performance.

As mentioned above, the performances of these sample desktop 3D printers have been assessed based on various parameters including build time, material usage and waste, and dimensional accuracy and surface roughness. However, different ranking methods allow one to rank an object based on the various criteria and to ignore the individual method, The results of Roberson et al. (2013) reveal the ranking of these machines with scores as follows:

1. Replicator (score: 0.31)
2. uPrint (score: 0.27)
3. SD300 Pro (score: 0.21)
4. 3D Touch (score: 0.14)
5. V-Flash (score: 0.06).

In several other case studies, experts approved the results of these selector systems, demonstrating their reliability, for instance in the case of a new design of mobile phone shell. For this model, material properties are not so important, but the tolerance on features can be relatively tight. The part's size is 90*40*15 mm. In addition, the model has a complex interior shape and precise feature geometry. Using the proposed selector system of Lan et al. (2005), the user can input values for selected parameters, some of which are numerical and some of which indicate a level of importance. When data were then put on the system, the feedback showed that the SLA process was the most appropriate technique in that case. The rest of the feasible alternatives were listed.

Another implementation example for the sand casting part, using a proposed selector system, is reported in Armillotta (2008). The results showed that among selected alternatives, the direct fabrication of a casting mold by SLS of sand has one of the highest scores, since it provides a good compromise of all the relevant criteria. The results also demonstrated for the rest of the alternatives that a LOM pattern would be an ideal choice due to the special suitability of the process to large thicknesses; a 3DP pattern would be faster to build and would keep material costs and the hourly rate of the RP system at a minimum; and an FDM pattern would allow a reasonable compromise among the advantages of the previous two solutions.

Likewise, the study of Mahapatra and Panda (2013) concluded that SLS (SLS 2500, printer model) is the most appropriate rapid prototyping (RP) system for better dimensional accuracy and surface quality, whereas 3DP (Z 402, printer model) is an appropriate RP system for a better build time. Consequently, manufacturers can efficiently select the most suitable techniques for their needs.

7.3 Implementing Additive Manufacturing

As an early study of the technology implementation process, Voss (1988) suggested a simple model consisting of pre-installation, installation and commissioning, and post-commissioning (Fig. 7.7). The implementation process should firstly consider the knowledge about the process and its interaction with the environment and other processes. Then, it should be concerned with the success and failure outcome of the process, and should recognize the changing definition of success over the lifecycle. Third, the factors influencing the implementation process and its success or failure should be identified. These factors may include organization, technical planning, business strategy, and operations management.

Small and Yasin (1997) extended this lifecycle model to an Advanced Manufacturing Technology (AMT) implementation process. They suggest the following for an effective implementation process:

1. Identifying and understanding the competitive global business environment
2. Strategic responses to this competitive demand, including AMT adoption and implementation planning
3. Establishing organizational goals and measuring performance during the implementation process
4. Adopting structures to meet organizational goals
5. Supporting the new technology structure by infrastructural adjustments
6. Evaluating the profitability of investment in the technology
7. Technology choice, and support for the adoption of the chosen system
8. AMT performance evaluation.

Consequently, studies and common sense suggest the importance of careful planning for the successful adoption of new technology. As a result several parameters should be understood before, during, and after the implementation process. In fact, the majority of the above-mentioned criteria are discussed in detail

Fig. 7.7 Lifecycle model of technology implementation. *Source* Voss 1988

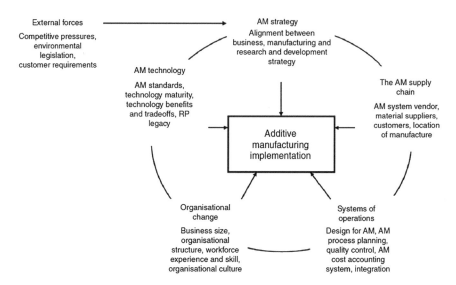

Fig. 7.8 AM implementation framework. *Source* Mellor et al. 2014

in Chap. 6 and Sects. 7.1, 7.2 and 7.3. However, organizational and structural changes, operational changes, and supply chain configurations will be discussed below.

Figure 7.8 is the typical framework that graphically explains the factors influencing the success of AM implementation. Factors are grouped into constructs along five areas, namely strategy, organization, supply chain, technology and operations (Mellor et al. 2014).

7.3.1 Organizational Changes

In many cases, poor organizational preparation for the adoption of emerging technology hinders the realization of potential benefits. For AM implementation, the decision to adopt should be accompanied by a change in jobs and functions, and thus a change in work practices and structure since AM technology offers unique manufacturing capabilities as well as new business models. As discussed in Sect. 4.1.2, when AM becomes a disruptive technology to conventional manufacturing, most organizations will find it difficult to adopt. Therefore, organizations must address the corresponding threats to prepare for future mandatory changes. Developing an integrated business plan based on corporate goals is an important step in the initiation of comprehensive organizational planning. This business plan can identify the contributions of each organizational unit, and each unit such as production, logistics, and marketing can prepare its own functional plans.

The size of an organization is also a critical factor in the technology implementation process. As discussed in Sect. 6.2.1, small and medium enterprises seem to benefit more from adopting AM technology. AM enables them to increase their competitiveness, find new markets, and acquire new customers (Khorram Niaki and Nonino 2017b). Currently, the majority of companies adopting AM belong to this classification. However, the implementation process for large companies seems to be faster. SMEs are more likely to encounter a lack of skills and trained employees, and require investment, whereas large companies can utilize many existing resources and expertise from across the firm(s) and also have more scope to invest. For instance, they are able to buy several AM machines. This may be useful for both covering the range of customer requirements as well as reducing the risks of machine breakdown, resulting in quicker demand fulfilment.

Organizational structure is also an important factor to consider in the successful implementation of a manufacturing technology, and those companies that initiate without redesigning structures may encounter difficulties. Studies on advanced manufacturing technology implementation usually prefer flatter and less complex organizational structures. These structures are more administratively decentralized and are more likely to succeed in a unique technology implementation, due to supervision that is more concentrated and increases individual responsibility. In other words, a balanced structure between high flexibility and control-orientation may facilitate greater success in AM implementation (Hopkinson and Dicknes 2003).

Before technology installation, companies should decide whether to retain and educate existing employees or employ new specialists. Technicians that are currently familiar with digital manufacturing systems (such as CNC machines) can, for example, be retained as they are more likely to be able to learn how to operate AM technologies. Of course, new expertise is required due to the new thinking required in the design process, drawing platforms, material properties, production process, and post-processing. Currently, this aspect may be a challenge to the implementation process due to the lack of a trained workforce. Despite the maturity level status of the technology, educational programs and training courses have not been well developed, so that the number of exiting specialist does not meet the demands of the workforce. To compensate for these gaps, AM machines currently play an important role in providing the information required in the preliminary steps of implementation. In addition, other external parties such as research institutions, consultancy services, software companies, and material suppliers can be chosen for collaboration.

In fact, an AM organization—at the minimum—needs special design engineers (with expertise in design for additive manufacturing), engineers for production, specialists for quality testing, adjusted customer relationship units, and preferably separate units for prototyping and manufacturing objectives while customer relationship activities need to be modified as customers may now be involved in almost all stages of new product development. Therefore, they should be able to conduct a systematic program to collaborate with their customers and understand their requirements.

Changing the technology of an organization needs adjusting knowledge or techniques that facilitate the inputs to outputs. Obviously, as with other new manufacturing technologies, formal education and training activities are an essential factor in successful adoption. AM organizations must invest significantly in skills development and education of cultural and technical aspects so that employees are able to recognize the full benefits and potential uses of the technology. These programs for AM technology should be carried out during the implementation process and post-commissioning. In particular, updating related knowledge is necessary due to the almost daily advances occurring in the technology in terms of new application environments, AM machines, materials, design software, and the novel techniques in additive manufacturing in general.

7.3.2 Operational Changes

Design

AM adopters should preferably undertake design in situ, as the design for AM will have a direct effect on productivity. It can contribute to post-processing, production planning and control, and satisfying customers' requirements. One of the objectives for DFAM (see Sect. 5.1.2 for detail) is the integration of components. Studies have proposed empirical guidelines for this purpose. For instance, Yang and Zhao (2015) proposed a consolidation method to reduce the part count and ease of assembly. It should be noted that here, part means a designed object that does not need an assembly operation, while assembly refers to a collection of two or more parts. In addition, the following principles should be considered in order to use the proposed method:

- The method must be suitable for the design with one material in assembly level rather than the part itself.
- The method should be appropriate for the design in multiple domains (namely stiffness, heat and dynamics).
- The method should consider the design for ease of process and better performance.
- The method has to be a process-based design methodology.
- The method should also consider operational aspects (namely cost, time, energy consumption, etc.) to reach a cost-effective design.

Figure 7.9 depicts the proposed design method for parts consolidation, in which the first layer contains the input data. The functional requirements are those related to what the part should do, while the performance requirements deal with how well the function is achieved. In addition, the conventional design file is also required as the initial input data. The consolidation processes begins using the given data dealing with part functions and required performances. The first step—called function integration—is the analysis of the given data and performing the consolidation at functional level. The second step is the optimization of design based on the other

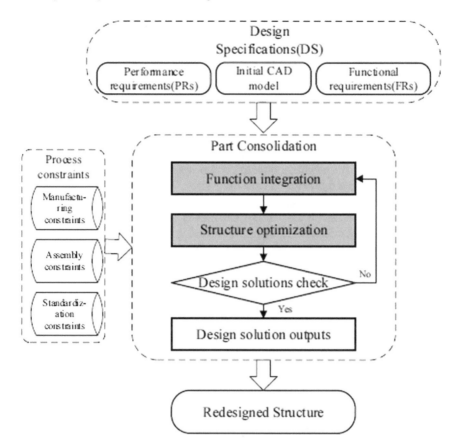

Fig. 7.9 Redesign method for parts consolidation. *Source* Yang and Zhao (2015)

capabilities of AM (such as reducing a part's weight). During these redesign steps the designer should also take account of some manufacturing, assembly, and standardization constraints. The steps would be repeated if there is no design solution, so that some modifications might be needed. Finally, the output is an optimized design with a lower part count and a higher performance.

Production planning and control

Advanced manufacturing technologies are the potentials for integrated production, with a single integrated system to control all the activities of a firm starting with raw materials and finishing with finished goods that are ready to be delivered to the customer. This production strategy leads to a reduction in the need for human interactions through maximizing the level of automation. In addition, it may reduce human errors, resulting in mistake proofing. In the case of AM, further mistake proofing are obtainable through automated post-processing, which usually depends on the skills of operators.

Since AM technology is based on a batch production system, companies should justify the tradeoff between maximizing the batch volume to obtain production efficiency, and/or shortening the production time to rapidly meet customers' incoming demands. In the latter case, the company needs to decide whether or not to produce even a single unit if the capacity is not fully utilized. This depends firstly on the quantity of machines that the company owns, to which they may assign a portion of facilities in order to rapidly fulfil urgent demands, while other facilities can be employed for full utilization. Secondly, the trade-off depends on market conditions, and a delay may result in loss of the order. Therefore, in this case, meeting demand as quickly as possible is the best production strategy even if it means losing on the optimum operational cost.

In addition, production planning and control for AM requires a choice to made concerning several parameters. These parameters are part orientation, optimized build volume, layer thickness, support structure generation, process speed, laser power (for laser sintering systems), hatching strategies, and temperature control. The following paragraphs explain these parameters.

Selecting part (build) orientation

A different part orientation in the AM process might directly affect mechanical properties, support structure, costs and quality. The optimum part orientation needs a detailed description, and some examples are presented below to help the reader understand the importance of this matter. For instance, in the SLS process, the Z-axis must be used as a reference where the tensile modulus has to be improved, while the reference has to move to XY for better elongation properties (Munguía et al. 2008). Generally, build height should be reduced if only printing a few units, the build chamber should be maximized if printing many units, and the build height should be reduced per unit.

Optimizing build chamber

Maximizing the build chamber is a target in order to reduce costs, since the incomplete use of the available chamber leads to inefficient machine operation. However, there is a limitation to this. For instance, in the SLS process, a 1–2 cm gap from each wall is necessary in order to avoid unsintered powder and imperfections.

The process starts with the selection of a part from the basket of parts, to insert in the platform. Then, the other components should be added to maximize the build volume. Figure 7.10a shows the suggested algorithm for packing build volume, and Fig. 7.10b depicts the sample maximized build chamber (Baumers et al. 2016).

Layering strategy

The selection of layer thickness contributes to both quality and cost. However, although it is possible to set a variable layer thickness in a build, most users set a fixed layer thickness. For instance, as reported in Munguia et al. (2008), for the SLS process, this is usually 0.1 mm and for DMLS from 20 to 60 mm is considered a suitable measure of thickness.

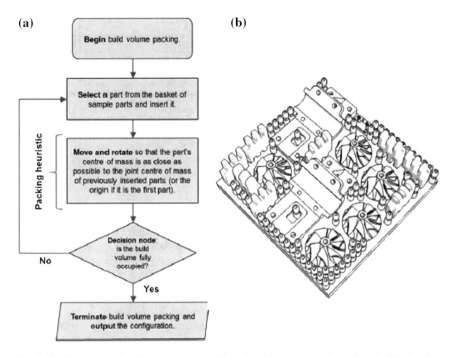

(a)

Begin build volume packing.

Select a part from the basket of sample parts and insert it.

Packing heuristic

Move and rotate so that the part's centre of mass is as close as possible to the joint centre of mass of previously inserted parts (or the origin if it is the first part).

Decision node: is the build volume fully occupied?

No

Yes

Terminate build volume packing and output the configuration.

(b)

Fig. 7.10 Maximizing build chamber **a** packing algorithm; **b** sample packed build chamber. *Source* Baumers et al. 2016

Support generation and minimization

During the process, the support material upholds the external and internal surfaces to keep a steady geometry with a structure using scaffolding. Designing the optimal support structure helps to reduce material usage and corresponding costs, since this piece should then be removed from the object. On the other hand, it affects the quality and reliability of the 3D printed part. Although there is a lot of software for generating the support structure, users generally use their own custom support designs. The support material can usually be cleaned during post-processing (as in IJP), or can be recycled during the finishing operations (for SLS and SLM systems).

The last four parameters deal with equipment calibration. These parameters are usually kept constant during planning of the AM process. Users may set the laser power in order to improve part flexibility, for instance laser power is set to 13 W (in SLS) for polyamide material in order to obtain a more flexible part. These tasks are evaluated based on their importance in production planning (Table 7.4).

After preparing the final design file, the operator has to prepare the AM machine with the required raw material (powder resin, polymer spool) as well as the production source (laser, printing head). Almost all the available AM technologies require some post-processing. This is mostly due to flaws (such as a stair-stepping effect of the layer-based process). However, the duration of this operation depends on the type of technology and the accuracy of the machine.

Table 7.4 Importance of AM process parameters for production planning. Adapted from Munguia et al. 2008

Priority	Parameters	Score (%)
1	Part orientation	85
2	Optimized build chamber	80
3	Layer thickness	75
4	Support structure generation	72.6
5	Process speed	60
6	Laser power	57.6
7	Hatching strategies	55
8	Temperature control	42.6

Process parameters optimization

Several studies proposed an optimization model of energy consumption according to the specific criteria of the product and process (e.g. Baumers et al. 2013; Paul and Anand 2012; Xu et al. 2015; and Kellens et al. 2010). AM process parameter selection is an important facet of sustainable manufacturing through enhancing productivity (time) and accuracy. The studies considered factors such as part geometry, build orientation, layer planning, type of apparatus, laser dose, and scanning speed to propose an optimization model for reducing time, cost, and energy consumption (energy consumption is actually a function of these factors). The influence of these parameters in various AM technologies is different. As an example of this, Xu et al. (2015) conducted experiments on the binder jetting system. They proposed a mathematical model to demonstrate the correlation between part geometries and energy consumption during production. The case was analyzed on a part that was 3D printed with a jetting system. This model provides a tool to optimize part geometry design with respect to energy consumption.

An experiment was performed on a cylinder (with radius of 1.5 mm and height of 4 mm), as seen in Fig. 7.11a, b. One of the important factors driving energy consumption is printing orientation. In this case, two angle values of printing orientation were considered: 0.00 and 90.00°. In this case, printing orientation refers to the angle between the build orientation vector (Z), and the normal vector of the top surface (nt). Figure 7.11c, d respectively show simulated power and experiment records for the two printing orientations, 0.00 and 90.00°. The first cylinder was printed in forty layers, each with a layer thickness of 100 μm and 4 mm printing height. The power curves of the printing process at 0.00° orientation are depicted in Fig. 7.11c, There are forty spikes on both simulated power curve and experimental records, which refer to the 3D printing of each layer. For the second cylinder, printed in thirty layers with a 3 mm printing head and a 90.00° printing orientation, the power curves are presented in Fig. 7.11d. The results in the figure demonstrate that the greater the printing height, the more process energy is consumed. Evidently, this is because the greater printing height needs a larger number of printing layers, and a correspondingly greater printing time. As for the part orientation, the process's energy consumption at 90.00° is 20.7% less than the energy consumption at a 0.00° printing orientation.

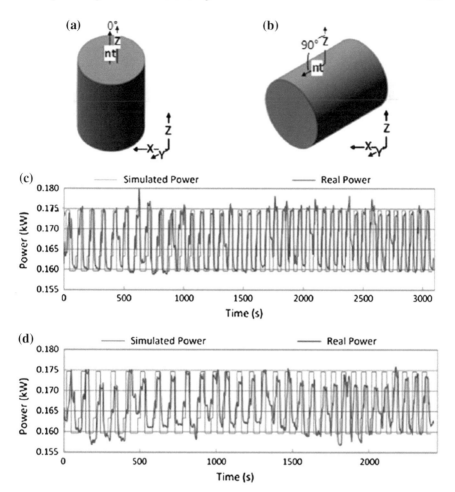

Fig. 7.11 3D printing power curves. *Source* Xu et al. (2015)

As mentioned before, the influences of these parameters in energy consumption will vary between the different AM technologies. The build time is the most important factor as the electrical power of each system is approximately constant during the printing, so the energy consumption is directly dependent on the build time. Consequently, to minimize energy consumption the build time must be first minimized. To do so in a jetting system for instance, the most important parameter is the height of the part, while in a FDM system minimizing the support structure is the most important factor. Figure 7.12 shows the level of influence of the parameters in different AM systems including the Thermojet (3D Systems Inc.), FDM 3000 (Stratasys), and a SLS system, the EOSINT M250 (EOS). For example, it can be seen that printing orientation has no influence on energy consumption when using FDM for manufacturing. Based on the study of Mognol et al. (2006) with a

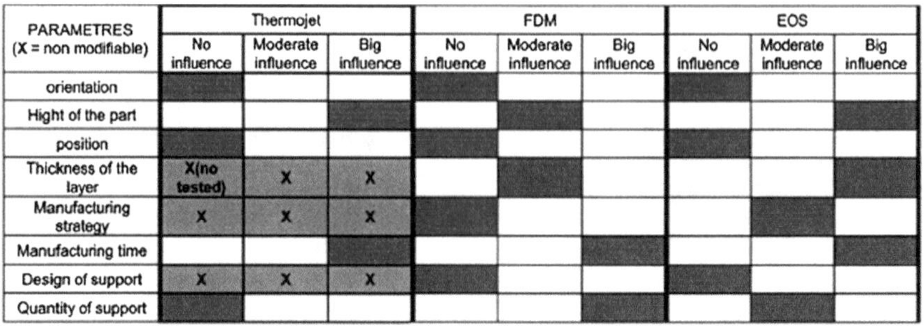

PARAMETRES (X = non modifiable)	Thermojet			FDM			EOS		
	No influence	Moderate influence	Big influence	No influence	Moderate influence	Big influence	No influence	Moderate influence	Big influence
orientation									
Hight of the part									
position									
Thickness of the layer	X(no tested)	X	X						
Manufacturing strategy	X	X	X						
Manufacturing time									
Design of support	X	X	X						
Quantity of support									

Fig. 7.12 Influences of the factors affecting in energy consumption. *Source* Mognol et al. (2006)

Table 7.5 The most used methods for quality control in AM processes. Adapted from Munguia et al. 2008

Quality control methods	(%)
Manual instruments inspection	66.7
Visual and tactile inspection	44.4
Coordinate Measuring Machine (CMM)	33.3
Mechanical tests	22.2
Microscopy inspection for surface	22.2
Random mechanical testing	11.1
Other methods	11.1

good set of parameters, it is possible to save 45% of electrical energy with the Thermojet, 61% with the FDM, and 43% with the EOS system.

Quality control

A quality management (QM) system should consider all the elements contributing to product development in order to ensure the quality of the final product. Different AM technologies need different quality plans. For instance, an electron beam melting (EBM) system needs a vacuum and higher temperature to relieve stress and porosity in order to minimize the need for post-processing while this is not the case for jetting systems.

Quality inspectors usually use different testing methods and instruments. These methods may involve visual inspections using manual instruments, and/or using measuring machines. Table 7.5 reports the results of a survey in which the most used quality control methods in AM are identified (Munguía et al. 2008). The majority of AM adopters (approximately 67%) use manual instruments for inspections such as testing dimensional measurements, and about 45% check the 3D printed parts through simple visual inspections.

The following describes the AM quality control system, using an example of the selective laser sintering (SLS) process from the study of Schmid and Levy (2014). This QM program includes five important elements of manufacturing, namely equipment, raw material, production process, batches, and the final 3D printed part.

Obviously, the important factor when considering the equipment relates to the initial process of selecting the appropriate technology and its suppliers. However, some activities dealing with maintenance also need to be conducted. For instance, daily-based maintenance can include cleaning, removing deposits on the laser window and so on. Other interval-based maintenance may include full services of lasers and temperature-checking devices, and replacement parts like filters. Machine suppliers usually provide these instructions in a service manual. As with conventional manufacturing equipment, careful consideration of these services is necessary in order to reach a higher level of product quality.

There is also another way to ensure the performance of the system. A reference part can be periodically built to check the parameters over time. These parameters may include surface finish, dimensional accuracy, tolerances, weight and density.

In terms of material, suppliers should guarantee the quality of raw materials, however, the producer needs to understand the controlling points and carry out its own control plan. For instance, several important criteria regarding quality should be considered when purchasing the powder-based plastic raw material. These parameters may include particle size of the powder, bulk density, and thermal properties (melting point and recrystallization point). In addition, some internal experiments can be done as well such as checking the thermal properties (can be done by differential scanning calorimetry) of plastic powders. Obviously, there are other specifications for other AM raw materials, from which a material quality plan can be extracted.

During the process, detailed documentation is needed in order to trace product quality issues. An automatic (or manual) recording system can be developed to document process parameters such as build orientation, laser power, temperature profiles, layer thickness, and scanning speed, etc. Some specialized production software (EOSTAT of EOS for example) can also assist in recording and maintaining these data. As the quality of product is linked to the process parameters, this documentation may help the producer to analyze the quality based on selected parameters.

A quality inspection for every single final part is required. These inspections may involve checking visual quality matters, controlling surface finish and part density, and controlling dimensions through use of optical instruments, etc. In addition, producers conduct several post-processing activities in order to reach a higher level of quality in the final 3D printed part. As discussed in Chap. 1, these finishing activities may include infiltration (which infiltrate the part with another material), surface optimization (vibratory grinding), and coating (e.g. color).

Generally, for each AM system, an Ishikawa diagram (fishbone) should be designed to identify the quality-related elements. Figure 7.13 illustrates an example of this diagram for the SLS process, which is adapted from the study of Schmid and Levy (2014).

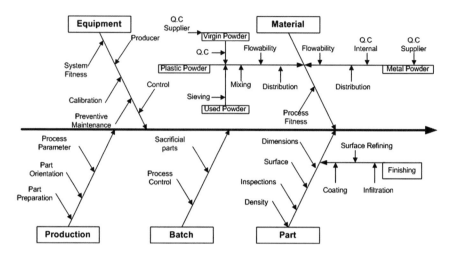

Fig. 7.13 Quality-related parameters of the SLS process

AM and IoT

The use of the internet is also having an influence on manufacturing. However, the contribution of the internet to manufacturing and production is still under study. The advent of the Internet of Things (IoT) paradigm has influenced all aspects of our life (smart cities, smart clothes, smart home, etc.) and production and manufacturing as well. The developing networks between objects and designs (such as online 3D printing platforms) can be seen as the emergence of the IoT in the way we produce and consume.

The following paragraphs provide a discussion on how manufacturing can be combined with the IoT in order to contribute to the creation of innovation. In other words, it suggests the stages of implementation of AM technology in order to realize smart manufacturing (Caputo et al. 2016).

First, the additively manufacturing parts should have a unique radio-frequency identification code (RFID). This code has great importance in the IoT for identifying products that are currently equipped with a RFID tag. However, in the case of AM technology, this code can be imbedded inside the layers of material and the addition of a RFID tag is therefore no longer needed. In this way, any 3D printed parts can directly participate in IoT chains. Second, items could be equipped with active sensors connected by a pre-defined code. The objects can be uniquely identified and are also equipped with sensors. AM machines can also be constantly connected to the internet so that the equipment can share data in the cloud.

3D printed parts and AM machines are uniquely identifiable on the network, are equipped with sensors, and are always connected to the internet. Consequently, the result of these facilities will be the useful flow of data. These data are receivable from both the product and the AM machine. Therefore, in this case, manufacturing

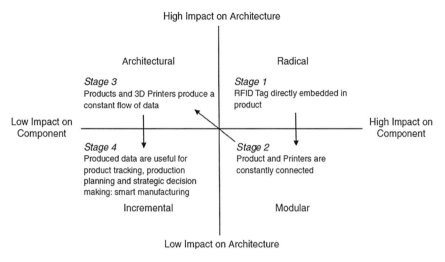

Fig. 7.14 Implementation of AM with the IoT. *Source* Caputo et al. 2016

can at least be operated and monitored remotely. Moreover, the existence of constant data as well as connected items may lead to smart manufacturing that can learn, adopt, and optimize the processes. AM technology can therefore be potentially considered as smart manufacturing, in the fourth stage of evolution where the designed information systems can contribute to all stages of production processes such as production planning, product tracking, fulfilling demand, process optimization and so on. Figure 7.14 illustrates the four evolutionary stages of smart manufacturing using AM technology accompanied with the IoT.

7.3.3 Supply Chain Changes

AM supply chain include material and equipment suppliers, manufacturers and customers. Machine suppliers also tend to be material suppliers and this is the current situation of the majority of vendors by which they are acquiring an exclusive market in order to guarantee their future markets. This was also the case with other emerging technologies where firms supplying the core of the technology also want to supply the supplements. However, they will partially lose exclusive power when the market is growing and the technology is maturing. AM suppliers can play an important role in the implementation process, given the status of the knowledge and maturity of the technology. They may provide several consultancy services advising on selecting AM systems and the appropriate materials for any given application.

From the customer's point of view, AM adopters should pursue the collaborative relationship with customers in the product development process in order to take full

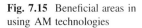

Fig. 7.15 Beneficial areas in using AM technologies

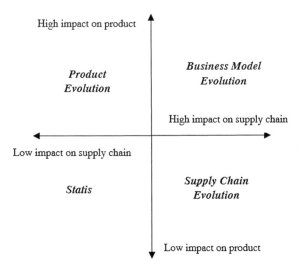

advantage of AM technologies. In addition, regarding the supply chain strategy and product characteristics, Cotteleer and Joyce (2014) described four main strategies that help guide adopters to their position. These impacts relate to the supply chain and product development (see Fig. 7.15).

The first strategy, called stasis, is the least risky path, in which companies improve value delivery by introducing AM to their existing business. AM provides them with benefits from using the technology for prototyping and the product development process to using it as a supplementary manufacturing method for the existing supply chain. The second strategy, namely supply chain evolution, uses the capability of AM to move the location of production closer to the users. Moreover, the on-demand style of production has a significant impact on the supply chain, enabling the company to manage risks and deal with unsold goods and inventory turnover.

In the third strategy, named product evolution, the impacts are mostly related to product development. Here, companies benefit from implementing AM technologies through producing innovative and highly complex products with advanced functionality and performance. More benefit still would come from manufacturing the customized products that deliver the size, shape, or color that each single customer expects. The last strategy is the business model evolution, where companies alter both product development and supply chain, seeking new business models. Because AM has the ability of mass customization, it provides for collaborative product development and production with open communities or customers, driving customer satisfaction upwards and enhancing innovativeness. Supply chain disintermediation refers to a flatter supply chain with a minimum of logistic and partners involved, as for instance with manufacturing at the point of consumers or prosumers.

Seven Key Facts

- There are three probable futures for AM: (1) product development will move to iterative and agile processes, (2) production of spare parts will occur in situ, (3) a larger share of products with multi-materials and/or embedded electronics.
- The decision criteria for selecting AM technologies are categorized by: technology, geometry, performance, economy, and productivity.
- It is fundamental to carefully plan the adoption of AM technology, as in many cases poor preparations hinder the realization of potential benefits.
- Organizational structures for the adoption of AM should preferably be flatter, administratively decentralized and less complex.
- Production planning and control for AM requires choosing several significant parameters (namely part orientation, optimized build volume, layer thickness, support structure generation, etc.).
- AM quality inspectors usually use different testing methods and instruments: visual inspections, manual instruments and/or measuring machines.

References

Armillotta, A. (2008). Selection of layered manufacturing techniques by an adaptive AHP decision model. *Robotics and Computer-Integrated Manufacturing, 24*(3), 450–461.

Baumers, M., Tuck, C., Wildman, R., Ashcroft, I., Rosamond, E., & Hague, R. (2013). Transparency built-in. *Journal of Industrial Ecology, 17*(3), 418–431.

Byun, H. S., Lee, K. H. (2005). A decision support system for the selection of a rapid prototyping process using the modified TOPSIS method. *International Journal of Advanced Manufacturing Technology, 26*(11–12), 1338–1347.

Caputo, A., Marzi, G., & Pellegrini, M. M. (2016). The Internet of Things in manufacturing innovation processes: Development and application of a conceptual framework. *Business Process Management Journal, 22*(2), 383–402.

Chung, C. W., Hong, J., Lee, A., Ramani, K., & Tomovic, M. (2003, January). Methodology for selection of rapid tooling process for manufacturing applications. In *ASME 2003 International Mechanical Engineering Congress and Exposition* (pp. 23–30). American Society of Mechanical Engineers.

Cotteleer, M., & Joyce, J. (2014). 3D opportunity: Additive manufacturing paths to performance, innovation, and growth. *Deloitte Review, 14,* 5–19.

Hopkinson, N., & Dicknes, P. (2003). Analysis of rapid manufacturing—Using layer manufacturing processes for production. *Proceedings of the Institution of Mechanical Engineers, Part C: Journal of Mechanical Engineering Science, 217*(1), 31–39.

Holmström, J., Holweg, M., Khajavi, S. H., & Partanen, J. (2016). The direct digital manufacturing (r)evolution: Definition of a research agenda. *Operations Management Research, 9*(1–2), 1–10.

Jiang, R., Kleer, R., & Piller, F. T. (2017). Predicting the future of additive manufacturing: A Delphi study on economic and societal implications of 3D printing for 2030. *Technological Forecasting and Social Change, 117,* 84–97.

Kellens, K., Yasa, E., Dewulf, W., & Duflou, J. (2010). Environmental assessment of selective laser melting and selective laser sintering. *Going green—Care innovation: From legal compliance to energy-efficient products and services, paper,* (2.14), 5.

Kengpol, A., & O'Brien, C. (2001). The development of a decision support tool for the selection of advanced technology to achieve rapid product development. *International Journal of Production Economics, 69*(2), 177–191.

Khorram Niaki, M., & Nonino, F. (2017a). Additive manufacturing management: A review and future research agenda. *International Journal of Production Research, 55*(5), 1419–1439.

Khorram Niaki, M., & Nonino, F. (2017b). Impact of additive manufacturing on business competitiveness: A multiple case study. *Journal of Manufacturing Technology Management, 28*(1), 56–74.

Khrais, S., Al-Hawari, T., & Al-Araidah, O. (2011). A fuzzy logic application for selecting layered manufacturing techniques. *Expert Systems with Applications, 38*(8), 10286–10291.

Lan, H. (2009). Web-based rapid prototyping and manufacturing systems: A review. *Computers in industry, 60*(9), 643–656.

Lan, H., Ding, Y., & Hong, J. (2005). Decision support system for rapid prototyping process selection through integration of fuzzy synthetic evaluation and an expert system. *International Journal of Production Research, 43*(1), 169–194.

Mahapatra, S. S., & Panda, B. N. (2013). Benchmarking of rapid prototyping systems using grey relational analysis. *International Journal of Services and Operations Management, 16*(4), 460–477.

Masood, S. H., & Soo, A. (2002). A rule based expert system for rapid prototyping system selection. *Robotics and Computer-Integrated Manufacturing, 18*(3), 267–274.

Mellor, S., Hao, L., & Zhang, D. (2014). Additive manufacturing: A framework for implementation. *International Journal of Production Economics, 149,* 194–201.

Mognol, P., Lepicart, D., & Perry, N. (2006). Rapid prototyping: Energy and environment in the spotlight. *Rapid prototyping journal, 12*(1), 26–34.

Munguía, J., de Ciurana, J., & Riba, C. (2008). Pursuing successful rapid manufacturing: A users' best-practices approach. *Rapid Prototyping Journal, 14*(3), 173–179.

Paul, R., & Anand, S. (2012). Process energy analysis and optimization in selective laser sintering. *Journal of manufacturing systems, 31*(4), 429–437.

Rao, R. V., & Padmanabhan, K. K. (2007). Rapid prototyping process selection using graph theory and matrix approach. *Journal of Materials Processing Technology, 194*(1), 81–88.

Roberson, D. A., Espalin, D., & Wicker, R. B. (2013). 3D printer selection: A decision-making evaluation and ranking model. *Virtual and Physical Prototyping, 8*(3), 201–212.

Schmid, M., & Levy, G. (2014). *Quality management and estimation of quality costs for additive manufacturing with SLS.* ETH-Zürich.

Small, M. H., & Yasin, M. M. (1997). Advanced manufacturing technology: Implementation policy and performance. *Journal of Operations Management, 15*(4), 349–370.

Voss, C. A. (1988). Implementation: A key issue in manufacturing technology: The need for a field of study. *Research Policy, 17*(2), 55–63.

Yang, S., & Zhao, Y. (2015). Additive manufacturing-enabled design theory and methodology: A critical review. *International Journal of Advanced Manufacturing Technology, 80.*

Xu, X., Meteyer, S., Perry, N., & Zhao, Y. F. (2015). Energy consumption model of Binder-jetting additive manufacturing processes. *International Journal of Production Research, 53*(23), 7005–7015.

Appendix A—Methodological Note

A.1 Survey Design

The survey questionnaire was sent to 105 key respondents (e.g. CEO, operation managers, R&D managers, product designer, business development managers, etc.) of companies around the world adopting AM technologies. In total, 105 companies from 23 countries participated in this survey research. Among which a number of global and well-known companies participated in this survey such as General Electric; General Motors; Airbus; Ford Motor Company; Lamborghini; Bell & Howell; Ducati Motor Holding; Valeo; Alcoa and Festo. Below the questionnaire.

© Springer International Publishing AG 2018   221
M. Khorram Niaki and F. Nonino, *The Management of Additive Manufacturing*,
Springer Series in Advanced Manufacturing,
https://doi.org/10.1007/978-3-319-56309-1

PART A
Firmographics questions

What is your company name?

What is your current position in the company?

In what country is your company currently headquartered?

In what year was your company founded?

In what year was your company implemented AM technologies?

About how many employees work at your company?
≤ 50 □ ≤ 250 □ > 250 □

How much was your company's revenue in the last year?
≤ 2 Million €□ ≤ 10 Million €□ ≤ 50 Million €□ > 50 Million €□

Which of the following best describe your AM implementation objectives?
Rapid Prototyping □ Rapid Manufacturing □ Rapid Tooling □

Which of the following categories best describes your products?
Consumer Products □ Industrial Products □ Art □ other (please specify....)

Which of the following industry best describes your AM application?
Automotive □ Aerospace □ Defense □ Medical □ Dental □ Education □ Industrial components □ Architecture □ Art □ Jewelry □ Research Institution □ other (please specify....)

Which type of material does your company use in additive manufacturing?
Plastic □ Metal □ Ceramic □ other (please specify....)

Which type of technology your company currently uses?
Selective Laser Sintering □ Direct Metal Laser Sintering □ Fused Deposition Modeling □ Stereolithography □ Electron Beam Melting □ Desktop 3D Printer □ Laminated Object Manufacturing □ Other (please specify....)

Which of the following dimensions best describes your product Size?
*Small [≤ 250*200*200 mm] □ Medium [≤ 500*400*400 mm] □ large [>500*400*400 mm] □*

Which of the following best describes the size of your production lot?
Small [≤ 40 parts] □ Medium [≤ 200 parts] □ large [>200 Parts] □

Which of the following phrases best describes geometric complexity of your product?
Very high □ High □ Average □ Low □ Very Low □

Which of the following best describe the trend of your products' demand?
Predictable □ Unpredictable

Which of the following phrases best describe your products?
Innovative □ Functional

PART B

Main questions

The following contains questions about your experience with additive manufacturing technology and its impacts on industrial worlds.

In comparison with the conventional manufacturing, to what extend do you perceive the effectiveness of AM in the following sentences?

Q 1- AM reduces waste.

Very Effective □   *Effective* □   *Neither Effective nor Ineffective* □   *Ineffective* □   *Very Ineffective* □

Q 2- AM rises the recycling rate of raw materials.

Very Effective □   *Effective* □   *Neither Effective nor Ineffective* □   *Ineffective* □   *Very Ineffective* □

Q 3- AM increase the Return on Investment.

Very Effective □   *Effective* □   *Neither Effective nor Ineffective* □   *Ineffective* □   *Very Ineffective* □

Q 4- AM reduces labor costs.

Very Effective □   *Effective* □   *Neither Effective nor Ineffective* □   *Ineffective* □   *Very Ineffective* □

Q 5- AM facilitates producing of the lightweight parts.

Very Effective □   *Effective* □   *Neither Effective nor Ineffective* □   *Ineffective* □   *Very Ineffective* □

Q 6- AM shrinks time to market of new product.

Very Effective □   *Effective* □   *Neither Effective nor Ineffective* □   *Ineffective* □   *Very Ineffective* □

Q 7- AM reduces energy consumption.

Very Effective □   *Effective* □   *Neither Effective nor Ineffective* □   *Ineffective* □   *Very Ineffective* □

Q 8- AM reduces environmental emissions.

Very Effective □   *Effective* □   *Neither Effective nor Ineffective* □   *Ineffective* □   *Very Ineffective* □

Q 9- AM reduces inventory turnover.

Very Effective □   *Effective* □   *Neither Effective nor Ineffective* □   *Ineffective* □   *Very Ineffective* □

Q 10- AM reduces processing times.

Very Effective □   *Effective* □   *Neither Effective nor Ineffective* □   *Ineffective* □   *Very Ineffective* □

Q 11- AM enables producing of high complex parts.
Very Effective □   *Effective* □   *Neither Effective nor Ineffective* □   *Ineffective* □
Very Ineffective □

Q 12- AM leads to better product's functionality and aesthetics.
Very Effective □   *Effective* □   *Neither Effective nor Ineffective* □   *Ineffective* □
Very Ineffective □

Q 13- AM produces parts with better dimensional accuracy.
Very Effective □   *Effective* □   *Neither Effective nor Ineffective* □   *Ineffective* □
Very Ineffective □

Q 14- AM enables producing full-customized parts.
Very Effective □   *Effective* □   *Neither Effective nor Ineffective* □   *Ineffective* □
Very Ineffective □

Q 15- AM improves our competitiveness in comparison with our competitor.
Very Effective □   *Effective* □   *Neither Effective nor Ineffective* □   *Ineffective* □
Very Ineffective □

Q 16- AM leads accessing to new markets.
Very Effective □   *Effective* □   *Neither Effective nor Ineffective* □   *Ineffective* □
Very Ineffective □

Q 17- AM enables acquiring also new customer in the same market.
Very Effective □   *Effective* □   *Neither Effective nor Ineffective* □   *Ineffective* □
Very Ineffective □

Q 18- AM assists innovativeness.
Very Effective □   *Effective* □   *Neither Effective nor Ineffective* □   *Ineffective* □
Very Ineffective □

Q 19- AM reduces occupational health hazard.
Very Effective □   *Effective* □   *Neither Effective nor Ineffective* □   *Ineffective* □
Very Ineffective □

In comparison with the conventional manufacturing, to what extend do you agree with the following sentences?

Q 20- AM requires raw materials that are more expensive.
Strongly disagree □   *Disagree* □   *Neither agree nor disagree* □   *Agree* □
Strongly agree □

Q 21- AM requires machines that are more expensive.
Strongly disagree □   *Disagree* □   *Neither agree nor disagree* □   *Agree* □
Strongly agree □

Q 22- A sufficient range of raw materials is available.
Strongly disagree □   *Disagree* □   *Neither agree nor disagree* □   *Agree* □
Strongly agree □

A.2 The Sample

Countries and industries

Respondent companies are established in major developed countries (75.2%), developed countries (17.1%), and developing countries (8%). Our sample shows relatively low diffusion level in developing economies. In fact, this is the case for all emerging technologies that may take long period to be introduced worldwide. However, considering the current affordable equipment, it is expected that in near future it will be widely applicable elsewhere. This categorization is based along the country classification of United Nation. Moreover, Fig. A.1 shows the countries of the sample firms, from which, USA, Germany, Italy and France constitute the most percentage of the AM adopters. The respondent is appropriate for the topic because it was composed of relevant and top level executives in the positions of chief executive officer (CEO), president, or vice president (40%), directors (23.8%), R&D, or design manager (16.2%), and other related professionals (20%).

Studies argued that the influence of Advanced Manufacturing Technology (AMT) on the companies in developing countries might vary in comparison with those belonging to developed countries (e.g., Ghani et al. 2002). In addition, researches revealed that the influence of technology on the business and organizational structure in a developing country was different from a developed country. AM can be considered as a strategic technology for creating value added parts and bringing back jobs for such a developed economics.

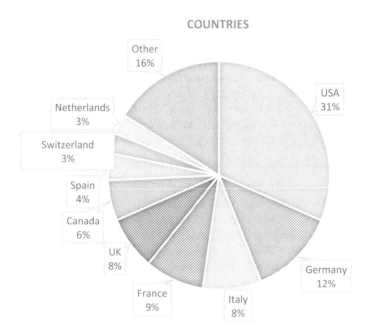

Fig. A.1 Countries in which companies are headquartered

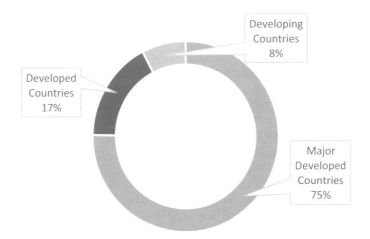

Fig. A.2 Distribution of respondents by countries' development level

As seen in Fig. A.2, our survey sample includes companies, belonging to major developed countries (75%), developed countries (17%), and developing countries (8%).

These companies works in/for a variety of AM application sectors and industries such as automotive, aerospace, marine, defense, electronics, medical, dental, education, architecture, art, jewelry, education and research institution, sporting goods, footwear, and food production. Since, some of them work as prototyping or manufacturing services for other company, they may contribute in various application sectors. Therefore, Fig. A.3 illustrates the cumulative percentage of each application sector. In which, three leading application sectors are those in aerospace industry and military, industrial components, and dental and medical components manufacturing. While, the frequency of using AM technologies for educational purposes (see Chap. 2 for detail) and research institutes shows interesting statistic, because the attraction for highly skilled academics and science center's shows potential capabilities of the technology for present and future.

AM systems and transition from conventional manufacturing

Figure A.4 illustrates the percentage of companies implementing a particular AM system. It is shown that the most widespread systems are Direct Metal Laser Sintering (DMLS), Fused Deposition Modeling (FDM), and Selective Laser Sintering (SLS). Despite the application of FDM, which is even affordable for consumer as well, using DMLS for additively manufacturing of parts with metal raw material depicts the maturity level of metal application in AM. In fact, plastic (photopolymer) material has been predominant in recent years, but currently using metal with AM seems also to be more attractive and effective manufacturing method. Moreover, other AM systems include jetting systems (binder jetting and polyJet), which is fully expected to leverage in industry and Digital Light Processing (DLP).

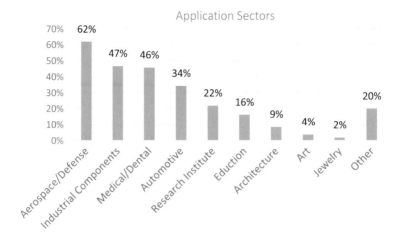

Fig. A.3 Distribution of respondents by application sectors

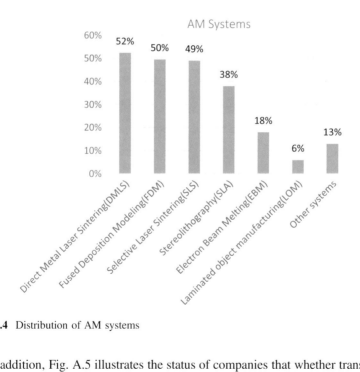

Fig. A.4 Distribution of AM systems

In addition, Fig. A.5 illustrates the status of companies that whether transitioned from conventional manufacturing to AM or start their business with AM. It shows that the majority of the companies adopted AM technologies to the existing manufacturing line or changed their manufacturing methods. Moreover, 28% of the companies work specifically with AM technologies. They are mostly 3D printing services, providing prototyping or manufacturing for third party or final consumers.

Fig. A.5 Percentage of companies, transitioned from conventional manufacturing to AM

Fig. A.6 Distribution of respondents by enterprise' size (*Small* employee, <50 and, turnover $\leq$ € 10 M€; *Medium* employee, <250, turnover $\leq$ € 50 M€; *Large* employee >250, turnover >€ 50 M€)

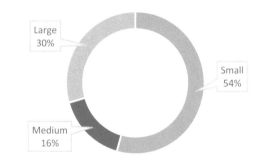

Fig. A.7 Distribution of respondents by company's experience

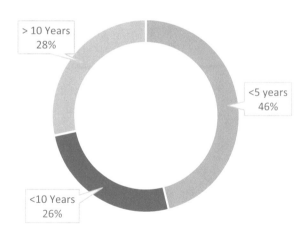

Size and experience of enterprises

Based on the European Commission's recommendation in 2003, and according to the revenue and number of employees, the sample includes small and medium enterprises (SMEs) (70.5%) and large companies (29.5%) (Fig. A.6).

As seen in Fig. A.7, our survey sample includes companies, which introduced AM technologies for less than five years (46%), ten years (26%), and more than ten years (28%).

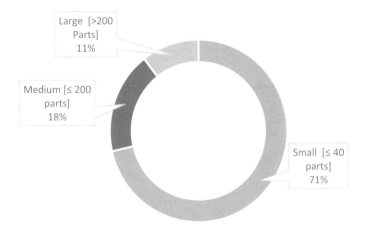

Fig. A.8 Distribution of respondents by production volume

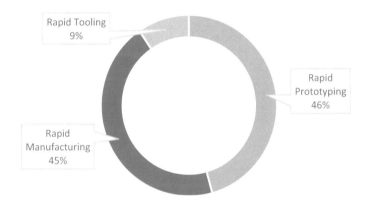

Fig. A.9 Distribution of respondents by implementation objective

Production volume and manufacturing objective

We adopted three categories for production volume. These categories include Small production volume [≤ 40 parts], Medium [≤ 200 parts], and Large [>200 parts]. As seen in Fig. A.8, our sample includes companies, which are 3D printing parts in small production volume (71%), medium (18%), and large ones (11%).

Generally, AM can be implemented for three main objectives such as prototyping, tooling and manufacturing. Rapid Manufacturing (RM) is defined as the use of AM technology to construct parts that are used directly as finished products or as components. Although the majority of the industries still use AM for Rapid Prototyping (RP) and Rapid Tooling (RT) to produce prototypes and tools for traditional manufacturing, currently AM machinery transcend its capability for manufacturing of end-usable parts.

Figure A.9 illustrates that our survey sample includes companies that their first priority is using AM for RM (46%), RP (45%), and RT (9%). Interesting result is

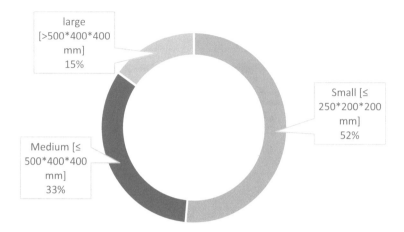

Fig. A.10 Distribution of respondents by product size

Fig. A.11 Distribution of respondent by material used

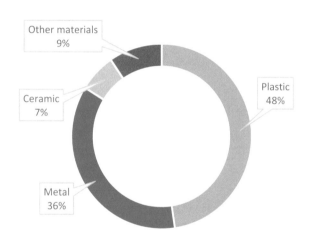

that companies using AM for manufacturing of end-usable products becomes in the same percentage of companies using it only for prototyping.

Product size and material typology

We categorized product size, as Small product size [$\leq 250 * 200 * 200$ mm], Medium [$\leq 500 * 400 * 400$ mm], and Large [>500 * 400 * 400 mm].

As seen in Fig. A.10, our survey sample includes companies, which are additively manufacturing small parts (52%), Medium (33%), and large ones (15%). This shows that the majority of respondents are using AM for small to medium product size.

Our survey sample includes companies that their first priority is using AM for plastic raw material (48%), metal material (36%), ceramic (7%), and other types of material (9%) (Fig. A.11).

Reference

Ghani, K. A., Jayabalan, V., & Sugumar, M. (2002). Impact of advanced manufacturing technology on organizational structure. *The Journal of High Technology Management Research, 13*(2), 157–175.

Glossary

3D Machine Vision (3D-MV) is a piece of scanning equipment that produces a three-dimensional scan of an object. This is a useful method for initiating reverse engineering, particularly in medicine. The combination of this technology with additive manufacturing will accelerate the product development process.

3D Printing (3DP) is the name of the technology invented by Massachusetts Institute of Technology and licensed to many companies. The technology uses a binder for solidification of powdered-based material. However, the phrase is also used instead of additive manufacturing.

Additive Manufacturing (AM) is the official and universal term for all technologies and their applications. According to the ASTM Standard F2792, it is defined as a process of joining materials to make objects from 3D model data, usually layer upon layer, as opposed to subtractive manufacturing technologies such as traditional machining.

Additive Process adds materials layer by layer. There are several techniques for this type of manufacturing.

Binder Jetting is a process that sprays a binder from an inkjet-style print head onto successive layers of powder. This is also a generic term for three-dimensional printing.

Break-even Analysis is a method used to identify when a business will be able to cover all its expenses and begin to make a profit. In this book, this method is used to finding the optimum production volume for AM, above which conventional manufacturing would be more efficient.

Bridge-tooling Process is the term used for rapid tooling. It is the process where moulded parts are produced at an early stage in the product development cycle, before the production tool is ready.

ColorJet Printing (CJP) involves two major components—core and binder. The system is suited to building full color concept models (i.e., architectural models and demonstration models). It can also be considered as a high-speed production system with a wide variety of materials and colors.

© Springer International Publishing AG 2018
M. Khorram Niaki and F. Nonino, *The Management of Additive Manufacturing*,
Springer Series in Advanced Manufacturing,
https://doi.org/10.1007/978-3-319-56309-1

Compound Annual Growth Rate (CAGR) is a specific business and investment term for the geometric progression ratio that provides a constant rate of return over the time period. During more than two decades of commercializing AM, the compound annual growth rate of worldwide revenues of all AM products and services was 25.4% in 2013. The growth rate has increased by 27.4% over the three-year-period from 2010 to 2012, reaching $2.2 billion in 2012.

Computer-Aided Design (CAD) is a system that uses computer to design objects and documentations, instead of manually drafting and drawing.

Design for Additive Manufacturing (DFAM) includes design guidelines that cover the capabilities of additive technologies. The new approach assists designers to gain the full advantages of this technology such as complex geometries, lightweight parts, integrated functionalities, and use of undercuts and hollow structures.

Design for Manufacturing (DFM) includes guidelines that cover conventional manufacturing, assisting designers in developing modular designs, using standard components, avoiding imperfections caused during manufacturing such as varying wall thickness, and sharp corners. These design attributes take into account conventional manufacturing constraints during the design process.

Design for Assembly (DFA) is a design attribute that focuses on ease of assembly.

Design for Disassembly (DFD) includes design strategies taking into account the future need for disassembling, repairing, or refurbishing a product.

Digital Light Processing (DLP) or DLP Projection is an additive technology that creates parts using a DLP projector to solidify the photocurable polymer on the base of a transparent vat.

Direct Digital Manufacturing (DDM) refers to the production of product using a set of technologies enabling the direct manufacture of parts from the digital data (or physical scan) without tooling and set-up. An original or copied file without any typesetting is sent to the manufacturing machine. In fact, this is such a generic term for additive technologies.

Direct Metal Laser Sintering (DMLS) is a laser-based additive technology that selectively fuses successive layers of a metal powder. The difference between DMLS and Selective Laser Sintering (SLS) is the production of metal parts without the need for a binder coating and the subsequent processing that would be required.

Eco-Indicator 95 is the European scale method, used for weighting the environmental impacts that damage ecosystems or human health. Eco-Indicator 95 contains 100 indicators for important materials and processes.

Electron Beam Melting (EBM) is an additive technology that melts powder-based material. EBM works with a thermionic emission gun, using a tungsten filament to make an electron beam. The process selectively melts metal powder in layers of 70–250 µm thickness with each layer preheated by scanning the beam to lightly sinter the particles. Arcam (Sweden) first commercialized the process in 1997. It is also known as Electron Beam Freeform Fabrication (EBF3) or Electron Beam Additive Manufacturing (EBAM).

Extrusion-based Process is a category of additive technology that creates parts by extruding the pre-heated material. Fused deposition modelling (trademarked by Stratasys) is one type of this process. The most popular material for this type of process is acrylonitrile butadiene styrene (ABS), which has properties that are similar to those of thermoplastics used in injection molding. The extrusion-based process is also known as the Thermal Extrusion method, Material Extrusion, or Melted and Extruded Modelling.

Flexible Manufacturing System (FMS) refers to a manufacturing system that can efficiently produce a variety of different components using the same resources at minimum changing cost. Additive manufacturing has extended the features of FMS and it can be considered as a novel and advanced FMS system.

Fused Deposition Modeling (FDM) is a kind of additive technology that creates parts by extruding and depositing heated material through a nozzle. FDM is known as a second extensively established AM system. It was invented by S. Scott Crump in Eden Prairie, Minnesota, and was commercialized and introduced by Stratasys in 1991. The available materials for this kind of system include investment casting wax, polycarbonates, polyphenyl-sulfon (PPSF), and most commonly acrylonitrile butadiene styrene (ABS). It is also known as Fused Filament Modeling (FFM), Fused Filament Fabrication (FFF), fused deposition process, or thermoplastic extrusion.

Industry 4.0 also known Industrie 4.0, refers to the developmental process in the management of manufacturing and production chains. Additive manufacturing is a part of this set of technologies. Other important components of this set of technology include the Internet of Things (IoT), cyber-physical systems, advanced robotics and cobots, edge and big data analytics and Augmented Reality (AR), etc. Industry 4.0 enables the so-called "smart factory".

Inkjet Printing (IJP) here refers to a kind of additive technology that prints three dimensionally and cures the photocurable resins. The process can print a number of acrylic-based photopolymers to create layers from printing heads containing many individual nozzles. There are several commercial machines for this system such as the PolyJet from Objet Systems, commercialized in 2000, and the InVision from 3D Systems, commercialized in 2003. The system is also known as MultiJet Printing (MJP), InkJet Photopolymer Printing, or Wax Deposition Modelling.

Jetting System also known as Material Jetting is a category of additive technology that creates parts by emitting a liquid from a printing head. It includes either direct or binder jetting. There are some commercial photopolymer jetting machines such as the PolyJet (which uses a combination of a photopolymer and jetting). It is also known as MultiJet Printing (MJP), or Inkjet Printing (IJP).

Laminated Object Manufacturing (LOM) is a kind of additive technology that creates parts through stacking layers of paper with a bonding material and cutting each layer of paper with a laser. Helisys first commercialized the process in 1991. In LOM, the material (i.e., polyvinyl chloride, PVC) is a special kind of paper that has a heat-sensitive adhesive applied to one of its sides. It is also known as Laminated Object Modelling.

Layer Manufacturing Technique (LMT) is a generic term for additive technologies, in which parts are fabricated by adding layer upon layer.

Liquid-based System is generic category of additive technologies that uses liquid material (a curable resin) to form solid objects. The process usually selectively cures regions of photosensitive polymers. This category of technology involves techniques such as Stereolithography (see SLA), and Jetting Systems.

Mass Customization usually uses a standard product as a base for customization and with the flexibility of individual customization. In mass customization, the production volume is the same as in mass production, but customers have options regarding the product designs. The products are usually modularized or bespoke for specific groups of end users. The use of mass customization can be found in a variety of industries (for instance, the electronic industries for PCs and cell phones).

Material Jetting is a category of additive technologies that creates parts by emitting a liquid from printing head. See also Jetting Systems.

New Product Development (NPD) is the process through which a new product(s) is launched. The process include eight stages: idea generation, idea screening, concept testing, marketing strategy, business analysis, product development, market testing, and commercializing.

Online 3D Printing Platforms seek to serve special needs dealing with AM for their customers. These platforms provides several services, some for the manufacturing of the object designed by consumers, while others offer (or host) designed files, and others cover both of these requirements.

Operations Management (OM) is about the way organizations produce goods and services. In other words, operations management is the activity of planning, organizing, and controlling all of the resources devoted to the production and delivery of a product or service. The resources include several items such as workforce, time, investment, technology, equipment, and information, etc.

PolyJet (a combination of a photopolymer and jetting) is trademark for the material jetting system of Stratasys that creates parts by spraying a liquid photopolymer from a print head. *See also Jetting system, and Material Jetting.*

Powder-bed Fusion is a category of additive technologies that creates parts using a heat source to selectively fuse powdered materials. Generally, any materials (for instance, polymers, metals, or ceramics) that can be melted and re-solidified can be used in this process. This category involves various additive technologies such as Selective Laser Sintering (SLS), Selective Laser Melting (SLM), Direct Metal Laser Sintering (DMLS), and Electron Beam Melting (EBM).

Prosumer (a combination of producer and consumer) generally refers to those consumers who produce their own requirements, or to consumers who directly participate in any value chain during production rather than being passive consumers. This concept can be found in Web 2.0 (i.e., Twitter, Facebook, and Instagram) where the consumer creates the value, or in online 3D printing platforms, where the consumer can be the designer or manufacturer.

Prototype generally refers to a first or preliminary physical version of a final product. Prototypes allow manufacturers to evaluate their design and even measure the performance of the products before mass production and distribution.

Rapid Manufacturing (RM) means the use of additive technologies to directly produce end-usable (finished) products or components from digital data.

Rapid Prototyping (RP) refers to the use of technologies to rapidly fabricate a part (prototype) before final production or commercializing. For many years additive technologies referred to rapid prototyping since it was used only for producing prototypes. Therefore, RP rather refers to the use of additive technologies to fabricate prototypes.

Rapid Tooling (RT) means the use of additive technologies to rapidly fabricate tools. These tools may include jigs and fixtures, molds and any type of complex manufacturing tools.

RepRap or replicating rapid prototype—is a home-use 3D printer, which can make a copy of itself or its components. Adrian Bowyer began the project in 2005, enabling AM machines to be owned and used by consumers at home. Although, at the start of this project the number of adopters was very small, users currently own more than 40,000 3D printers. Moreover, a range of different RepRap designs can be freely downloaded from its website.

Selective Laser Sintering (SLS) is a kind of additive technology that creates parts through selective sintering of powdered-material using a laser. Sintering is a process that heats the outsides of powder particles in order to fuse them together. The laser selectively scans the surface of a powder bed to create each layer. A variety of raw materials including polymers, ceramics and metals is currently available for the SLS process.

Solid Freeform Fabrication is a generic term also used in additive manufacturing. The term refers to the reduced complexity of additive manufacturing methods, allowing freeform design and fabrication of final product.

Solid-based System is a set of additive technologies that uses solid material to create parts. These systems include several additive technologies such as Fused Deposition Modelling (FDM), and Laminated Object Manufacturing (LOM).

Specific Energy Consumption (SEC) is defined as the energy consumption in the production of a material unit. The SI unit for specific energy is the joule per kilogram (J/kg).

Stereolithography (SLA) is a kind of additive technology that creates parts through curing and solidifying the selected portion of photocurable liquid resins, using an ultra-violet laser on the platform. The platform is then lowered, and a fresh layer of liquid resin is deposited over the previous layer. It was the first additive technology, commercialized by 3D Systems in 1987.

STL (STereoLithography interface format) is the file format for digitally defining the surface of a 3D object using a series of triangular facets. It translates computer-aided design (CAD) to a file readable by AM machines. The format was first created by 3D Systems in 1987, and is currently the commonest format for additive manufacturing.

Supply Chain Management (SCM) is concerned with the holistic management of the supply chain as a whole, from suppliers to customers. According to Slack et al. (1998), SCM "is the management of the relationships and flows between string of operations and processes that produce value in the form of products and services to the ultimate consumer".

Support Structure is like a scaffold that is added to the object during additive manufacturing processes to prevent overhanging or disconnected parts falling away.

Three-Dimensional Printing (3DP) *see* 3D Printing.

Vat Photo Polymerization is a category of additive technologies that create parts through selectively solidifying a vat of liquid photo polymer by means of a laser beam or other light sources. This category involves different additive technologies such as Stereolithography (SLA), and Digital Light Processing (DLP) projection.

Uncited References

3D Hubs. https://www.3dhubs.com/.

3D Systems Inc. https://www.3dsystems.com/.

3T RPD LTD. https://www.3trpd.co.uk/.

ARCAM AB. http://www.arcam.com/.

ASTM F2792-10e1 Standard terminology for additive manufacturing technologies. ASTM International. http://enterprise.astm.org/filtrexx40.cgi?+REDLINE_PAGES/F2792.htm.

Cubic Technologies. http://www.cubictechnologies.com/.

DWS Systems. http://www.dwssystems.com/.

EnvisionTEC. https://envisiontec.com/.

EOS GmbH. https://www.eos.info/en.

Hornberger, L. E. (1993). *Rapid prototyping program*. California: Santa Clara University.

KOR Ecologic Inc. https://korecologic.com/.

Makerbot 3d printing company. https://www.makerbot.com/.

McKinsey Global Institute. http://www.mckinsey.com/.

NASA Jet Propulsion Laboratory (JPL), https://www.jpl.nasa.gov/.

Niaki, M. K., Nonino, F., Komijan, A. R., & Dehghani, M. (2017). Food production in batch manufacturing systems withmultiple shared-common resources: a scheduling model and its application in the yoghurt industry. *International Journalof Services and Operations Management, 27*(3), 345–365.

PwC (PricewaterhouseCoopers). (2014) "As 3-D printers become faster, easier to use, handle multiple materials, and print active components or systems, they will find use beyond rapid prototyping". Retrieved from http://www.pwc.com/us/en/technology-forecast/2014/3d-printing/features/future-3d-printing.html.

Sells, E. A. (2009). *Towards a self-manufacturing rapid prototyping machine* (Doctoral dissertation). University of Bath.

SmarTech Publishing. https://www.smartechpublishing.com/.

© Springer International Publishing AG 2018   239
M. Khorram Niaki and F. Nonino, *The Management of Additive Manufacturing*,
Springer Series in Advanced Manufacturing,
https://doi.org/10.1007/978-3-319-56309-1

Index

© Springer International Publishing AG 2018
M. Khorram Niaki and F. Nonino, *The Management of Additive Manufacturing*,
Springer Series in Advanced Manufacturing,
https://doi.org/10.1007/978-3-319-56309-1